Study Guide to Accompany

Chemistry
& Chemical Reactivity

SECOND EDITION

Kotz & Purcell

Harry E. Pence

State University of New York at Oneonta

Saunders College Publishing
Philadelphia   Fort Worth   Chicago   San Francisco
London   Tokyo   Sydney

Kotz and Purcell: Study Guide to accompany
                    CHEMISTRY AND CHEMICAL REACTIVITY, 2/E

ISBN # 0-03-053489-5

123 084 987654321

TABLE OF CONTENTS

ACKNOWLEDGEMENTS

Many individuals have helped me during the development of this study guide, and I'm happy to take this opportunity to express my appreciation to them. Jack Kotz generates enthusiasm in everyone whom he encounters, and it was a pleasure to work on this project with him. I wish to thank John Vondeling and Kate Pachuta for their patience and assistance during the writing of the first edition and Sandi Kiselica for her work on the second edition. Susan Harman deserves special mention for the invaluable contribution she has made to the second edition.

It's impossible to acknowledge all of my colleagues and former teachers who have contributed to this endeavor, but I do wish to specifically thank Larry Armstrong and Bruce Knauer. A special note of gratitude is necessary for the many students who have helped me to better understand the difficulties of learning chemistry.

Finally, I wish to express my deepest appreciation to those whose encouragement, assistance, and unfailing support have been an essential component in the completion of both the first and second editions of this book, my parents, my children, and most especially, my wife, Virginia.

PREFACE

This study guide/workbook is intended to serve as a supplement to the second edition of *Chemistry & Chemical Reactivity* by John C. Kotz and Keith F. Purcell. The organization is basically the same as in the first edition of this guide; each chapter coincides with a chapter in Kotz and Purcell's textbook. Thus, when a student is working on Compounds and Molecules, Chapter 3 in Kotz and Purcell, he or she should consult Chapter 3 of this book. Since the study guide supplements the textbook and class notes, studying will be most effective if all three are used in combination.

As in the first edition, each chapter in the study guide provides a list of Learning Goals and Important New Terms to help students to understand the material in the corresponding chapter in the textbook. In addition, most chapters include study hints that identify some of the more common student misunderstandings and mistakes.

Self testing is widely recognized to be an essential part of any study program. It both reviews what is already known as well as identifies topics that require further work. This guide is organized to encourage a systematic approach to self testing. Each chapter includes a Concept Test, which is intended to evaluate the student's knowledge of the definitions and simple concepts from the chapter. Since these concepts are the building blocks for more complicated material, they should be mastered first.

This study guide provides two further types of self testing. Practice Problems, complete with worked solutions, should provide the student with a good preliminary measure of his or her skills. The practice problems also refer to specific learning goals, so that students spend more time on the topics that seem to be causing difficulties.

Finally, the Practice Test gives the student a chance to work under conditions that are as close as possible to those of a real examination.

One of the more helpful components of this study guide should be the chapter entitled "Some Suggestions for the Student." This consists of practical suggestions that should make studying more effective and improve the student's performance in the course. Students are urged to read this section carefully.

This study guide also includes two chapters that do not have counterpart sections in the textbook by Kotz and Purcell. These are labeled "Special Sections" and deal with the use of equivalents and normality for titration calculations. Instructors who follow this approach in their classes should refer their students to this material.

SOME SUGGESTIONS FOR
THE STUDENT

You have probably purchased this book because you are a little worried about taking General Chemistry or perhaps have already gone far enough in the course to recognize that you're having problems. Don't be discouraged; even though you may have heard a great deal about how hard chemistry is, you can still not only pass but even obtain a good grade if you approach the course systematically. This guide is intended to help you organize your work.

Many students find Chemistry to be difficult because it requires the mastery of a number of different skills. You must be able to take good notes, organize your work, do mathematical calculations, develop good study habits, do some memorization, and most important, be able to translate problems from word statements into mathematical equations. This isn't as complicated as it sounds, but it does take some practice. Some students who have a good high school background may be able to get by for a while on what they already know, but eventually each individual will have to develop these skills in order to be successful. This chapter contains some suggestions that should help you to develop the skills needed to do well in this course.

There are probably more recommendations in this chapter than any student will wish to apply all at once, so pick a few now, then come back later for more ideas. In order to make this section as useful as possible, the ideas are arranged by topic with individual suggestions indicated by a check mark. Thus if you find that you're having trouble with taking tests, you may wish to go directly to that section and look for suggestions that appear to be useful. The important thing is to come back to this material several times in the coming semester and look for new ways to improve your skills.

LECTURE PREPARATION AND TAKING NOTES

✓ *Read about the material in your textbook before hearing it explained in lecture.*

Don't expect to understand everything in the chapter completely but just try to become familiar with some of the ideas and identify what topics seem hardest to grasp. When you go to the lecture, pay very close attention when these more difficult topics are discussed. If you have had high school chemistry, some of the material in the text will seem familiar. Don't let this give you a false sense of security. Your instructor in college may want some material done differently from the way you learned in high school. He or she has a good reason for wanting this, and if you try to continue to do things the old way, you may find yourself in trouble.

✓ *Make your notes as accurate and complete as possible.*

As you listen to the lecture, don't try to write down every word, but instead attempt to determine what the instructor thinks is important and make sure that your notes indicate these priorities.

If the instructor is describing a multi-step problem, he or she will usually indicate the proper order in which you should do each step of the procedure. This is usually an important part of the process, and so your notes should record that order.

✓ *Review and expand your notes after class is over.*

As soon as you can after each class, review the notes. Reading what you wrote down in class will usually remind you of words and phrases that you didn't have time to include. Add this material to your notes. Be sure to go over your notes as soon as you can, for your memory of these extra comments will soon fade.

Try writing notes only on the right hand page in your notebook, leaving the left page empty. As you read the notes, identify those sections that aren't completely clear. Then try to write a question that focuses on each idea you didn't understand and write it on the left hand page next to the confusing section. When you have a chance, ask your lecturer or teaching assistant these questions or find the answer in the textbook and write the answers in the appropriate place on the left-hand pages.

Once you have a good set of notes, it is time to really read the textbook carefully. If you have trouble with a certain problem type, copy an example problem from your text or study guide into the space on the left-hand page of your notebook. Also add any comments or special data that may be useful. In this way you will create a set of notes that includes supplemental information just where you need it most. Some students rewrite their notes and produce a beautiful notebook. The procedure suggested here may not produce a set of notes that looks as elegant as rewritten notes, but you will probably invest less time and your notes will be a more clear and complete study aid.

✓ *Don't be afraid to ask questions.*
When you can't understand a topic on your own, ask your instructor or teaching assistant for help. Students rapidly learn that the questions they fail to answer when studying often appear on examinations. Try to make sure that your questions are clearly stated and asked at an appropriate time. Whenever possible, think carefully about your question and make it as specific as possible. Vague questions usually produce answers that aren't very useful.

✓ *When studying, emphasize the material you don't understand.*

As you continue to study the material in a chapter, you should place less emphasis upon the topics that have already been mastered. Unfortunately it's a very human temptation to spend most of your time on the material that you already know. Avoid this by first identifying the topics that need the most work, then making a study plan that will focus on these topics.

✓ *Identify your weaknesses with self testing.*

The study guide has been designed to encourage this type of self-testing, but you can also write your own tests using worked examples from the textbook. Self-testing means that you try to duplicate the conditions that you will encounter on a real examination. Put away all your books and notes, then try to work the test without any help. This will not only reinforce what you do know but also help you to identify the material that needs more work.

✓ *Consider studying with other students.*

Many students find that it's helpful to study at least part of the time with one or two friends. Sometimes another student understands your problems better than an instructor, and often you can solidify your knowledge by explaining a concept to another student. This type of exchange can be of benefit to everyone.

USING THE CALCULATOR

You will probably be using a personal electronic calculator in this course. The modern calculator is not only very helpful but also very rugged. There are, however, some steps you can take to make sure that your calculator gives you the best possible service.

✓ *Really learn how to use your calculator.*

Practice using the calculator as much as

possible before the first examination. Read the instruction book and try to learn how to use special keys like 1/x and the parentheses that can be extremely useful.

✓ *Take good care of your calculator.*
Although the electronic calculator is quite rugged, it still should be treated with some care. Don't pile books on top of it in your back pack or drop it on the floor. Keep it dry and cool. If you're not sure of how much longer the batteries will last, be sure to have spares when taking an examination.

HOW TO USE THIS STUDY GUIDE

Each component in the Guide has been written with the idea that the best way to study chemistry is to test yourself frequently under the same conditions you will face on the examinations. Put away your textbook and notes and answer questions as though you were working on a test. Self-testing is the best way to reinforce what you already know as well as a good way to discover what topics need more study.

✓ *Make sure that you understand the concepts and terms.*
Once you have studied your notes and the chapter in the text, review the Learning Goals and Important New Terms in the study guide. Do you know all of the terms? Do you think you have mastered each skill described in the learning goals? If so, continue to the concept test. The concept test consists of simple questions on the basic ideas presented in the chapter. Although it doesn't include all the terms defined in the chapter, if you have trouble with the concept test it suggests that more study is needed. If you have trouble with these basics you will probably also have difficulty solving the problems.

✓ *Now you are ready to attempt the practice problems.*
In each case the solution is provided, but

don't use it unless you are completely unable to do the problem on your own. Each practice problem is keyed to one of the learning goals. For example, a boldface **L3** after a problem number indicates that it is based on learning goal 3 in the current chapter. The list of learning goals indicates what section of the textbook chapter you should review. At this time you may also wish to review the Study Hints to check for common mistakes that you may be making.

✓ *Next attempt the practice test.*
   When you have satisfied yourself that you understand all of the learning goals, proceed to the practice test. The answers of these questions are also related to the learning goals so that you can identify weak areas that you have not yet corrected. Work the practice test on a separate piece of paper, so that you can reuse these questions later when you're reviewing for the final examination.

IMPROVING PROBLEM-SOLVING SKILLS

   A major activity in any general chemistry course is solving numerical problems. Since many students have difficulty with this skill, it is an important skill for students to master.

✓ Try to understand the basic concepts <u>before</u> working the problems.
   If you don't really understand the basic concepts, it's tempting to just memorize the procedures in some of the worked examples and then try to apply what you've memorized like recipes from a cookbook. You may be able to work simple problems using this approach, but there are too many problem types and variations for you to memorize all of the recipes that you will need.

✓ Practice identifying the principle that is the basis of the problem.
   Recognizing the key principle which is the basis of a problem is the first and probably

the most important step in problem solving. Look for the key words or phrases in the problem that relate to the relationship that is being used. One useful way to practice the recognition of the possible problem types is to use a copying machine to make a copy of all the practice problems at the end of the chapter. Cut out each individual problem, and mix up the resulting scraps of paper. Then try to organize these problems by type, assigning each one to the appropriate learning goal from the study guide. Where it's appropriate, be sure that you know the mathematical relationship or formula required for each learning goal.

Worked examples in the text or guide can be very helpful. Cover the solution and try to set up how you would solve the problem. If you find you are completely stuck, uncover part of the solution for a hint. This kind of practice is especially important when you are studying for the final exam and trying to organize many different problem types.

✓ *Use a systematic approach to problem identification.*
There has been a great deal of research on problem solving techniques, and your instructor may well have some suggestions to help you. The method described in this section is based in part on an excellent book by G. Polya* and is suggested as a model of how you should attack problems.

First, read the problem carefully and make sure that you understand the question. Clearly identify what data is provided and especially the quantity which you are asked to calculate. Time is important on an examination, but you can waste time by misunderstanding the question and starting to work the wrong problem. Try to organize the

---

* Polya, G. How to Solve It (2nd Edition), 1971, Princeton University Press, Princeton, NJ

information to determine what is important and what is unnecessary. Some students circle the important points in the problem statement or make a rough drawing to help visualize the situation better. A data table is another good way to organize the information, and you will find this method used frequently in the text and study guide, especially when the problems become more complicated.

Second, when you are sure that you understand the problem, develop a plan for solving it. Try to relate this situation to what you already know. Look carefully at the data and the unknown; have you worked a similar problem before? In many cases, a problem that looks difficult is a combination of several easy problems, so try to break it down into simpler components. If you don't have enough information to solve directly for the unknown, can you combine what is available to obtain the missing data? Try to reason backwards; what data would you require to determine the requested unknown? Can you rearrange the available data to produce what you need? If you are totally unable to see how to work a problem, skip it for the time being. Perhaps coming back to the problem later will give you a fresh perspective.

Third, once you have devised a plan, execute it carefully. Don't try to skip steps or do too much in your head. This not only increases the chance of error but also decreases the possibility that you may see your mistake when you check the problem. Students frequently complain that they make "stupid mistakes." When your solution is presented in a clear, step-by-step way, it is easier to avoid simple mistakes as well as to find your errors.

Fourth, check your work if time permits, but even if you are rushed, always take a few seconds to ask yourself if your answer seems reasonable for this type of question.

Remember that the instructor may sometimes ask questions that have answers beyond your previous experience, so that every answer that looks unusual isn't necessarily wrong, but often the answer that looks strange reflects a mistake in your solution.

TAKING EXAMINATIONS

✓ *Be sure to get enough sleep the night before the exam.*

Being rested and knowing that you are well prepared should minimize the possibility of text anxiety during the examination. If you are already tired when you start the exam, you won't be able to think clearly and will be more likely to panic. The all-night study session is a well-established college tradition, but it has hurt far more students than it has helped.

✓ *Work carefully and systematically.*

Don't forget to use a systematic approach to problem solving like that described above. Try to distribute your time so that you start with the easier problems. If you become stuck on a hard problem, don't waste all of your time on it. Go on to other questions and come back to the difficult one later if possible. To decrease tension, stop from time to time, look away from your paper, and relax a few seconds. Be sure to use the right number of significant figures and include units with your calculations.

✓ *Learn from your mistakes.*

When your graded test is returned, don't just look at the grade and celebrate (or curse). Go over each question that you missed and try to understand what you did wrong so that you may be able to avoid that error on future examinations. If you find that you are making certain types of mistakes often, then you should check for that error when working tests and quizzes. Often the final exam is cumulative in chemistry courses. If that is the case in your course, you may see similar problems again.

Some Last Comments

Even if you do everything suggested, you may not find Chemistry to be an easy subject; many people don't. The goal is to make it less difficult and easier to understand. If this study guide helps to achieve this, it will have accomplished its purpose. Good luck!

CHAPTER 1
AN INTRODUCTION

LEARNING GOALS:

1. Distinguish between substances and mixtures, atoms and elements, chemical and physical properties, and intensive and extensive properties. (Sec. 1.3)

2. Convert lengths, masses, and liquid volumes from one metric (SI) unit to another. Also remember the important relationships that 1.000 Liter is defined as 1000 cm^3 and that one milliliter of water is usually considered to have a mass of one gram. The latter relationship is only exactly true at about $4°C$ but is often assumed in the absence of more precise information. (Sec. 1.4)

3. It's important to recognize the approximate relationship between the commonly used non-metric units and metric (SI) units, since this type of numerical conversion may be required. Also be able to express large and small quantities in correct scientific notation. (Sec. 1.4)

4. Understand the relationship between mass, volume, and density. Can you determine whether each of these is an extensive or an intensive property?

5. Know how to convert from one to another of the three widely used temperature scales, Fahrenheit, Celsius, and Kelvin.

6. Know the guidelines in the textbook for the use of significant figures. Especially notice the difference between the rule for addition and subtraction and that for multiplication and division. Be sure to use dimensional analysis when solving problems. (Sec. 1.5 & 1.6)

7. Understand the difference between precision and accuracy. (Sec. 1.5)

8. Review the idea of percentage and be able to use

this concept when doing calculations. (1.6)

IMPORTANT NEW TERMS:
(Following each term is the section of Kotz and
Purcell where the term is originally discussed.)

absolute zero (1.4)
accuracy (1.5)
atom(1.3)
Celsius degrees (1.4)
chemical property (1.3)
compound(1.3)
density (1.4)
dimensional analysis (1.6)
element (1.3)
extensive property (1.3)
Fahrenheit degrees (1.4)
gas (1.3)
heterogeneous (1.3)
homogeneous (1.3)
hypothesis (1.2
intensive property (1.3)
kelvin (1.4)
kilogram (1.4)
law (1.2)
liquid (1.3)
liter (1.4)
mass (1.4)
matter (1.3)
meter (1.4)
mixture (1.3)
molecule (1.3)
percent (1.6)
phase (1.3)
physical property (1.3)
precision (1.5)
second (1.4)
SI units (1.4)
significant figure (1.5)
solid (1.3)
solute (1.3)
solution (1.3)
solvent (1.3)
state (1.3)
substance (1.3)
temperature (1.4)

theory (1.2)
weight (1.4)

CONCEPT TEST

1. _____ is defined as anything that has mass and occupies space

2. A(n) _____ is the smallest particle of an element that retains the chemical properties of the element.

3. A compound is composed of identifiable units containing two or more different _____ chemically combined in specific ratios.

4. If sugar is dissolved in water to form a solution, the water is called the _____ and the sugar is called the _____

5. Properties that matter exhibits when it undergoes a change in atom arrangements or atom ratios are called _____ properties.

6. What is the mass in kilograms of 1.00 liter of pure water? _____.

7. For each of the properties listed, state whether it is an extensive or intensive property.

a. boiling point _____

b. density _____

c. mass   _____

d. temperature _____

8. What is the volume in cubic centimeters of a 1.00 liter volume? _____

9. _____ is the value of absolute zero in units of Celsius degrees.

10. The boiling point of water is _____ degrees on the Celsius temperature scale and is _____

degrees on the Fahrenheit scale.

11. Complete the table below by listing the appropriate numerical value represented by each metric prefix:

a. milli _____   b. centi _____

c. kilo _____   d. deci _____

e. micro _____   f. pico _____

12. Make the conversions indicated.

a. 2.05 L   = _____ mL

b. 0.12 km   = _____ m

c. 44.5 cm   = _____ mm

d. 450. mg   = _____ g

e. 22,500 cm^3 = _____ L

f. 235 mL of water weighs _____ g

13. Indicate the correct number of significant figures in the answer to each calculation.

a. 25.60/2.1 =   b. 5670 x 3.51 =

c. $\dfrac{300. + 56.8}{5}$ =   d. 2.01 + 0.03582 =

14. Round each of the following numbers to two significant figures and report the number in correct scientific notation.

a. 273.15 _____   b. 345,000,000 _____

c. 0.00735 _____   d. 0.00055 _____

15. Classify each of the following as either a physical or a chemical change.

a. a nail rusts   _____

b. antifreeze boils   _____

c. gasoline burns   _____

d. sugar dissolves   _____

STUDY HINTS:

1. As in the textbook, the problems in this study guide are worked with an electronic calculator, and the answer is rounded to the correct number of significant figures only at the end of the problem. Rounding off intermediate values may produce slightly different results from those in the text.

2. Students sometimes underestimate the importance of density. It is a simple concept, but one that will be important for many problems later in the course. Be sure to remember it.

3. Most of the unit conversions you will do in this course will involve only metric units. Remember the prefix indicates the magnitude of the conversion factor, and dimensional analysis will help you to decide whether to divide or multiply by the conversion factor.

4. Remember that all numerical answers should be obtained by carrying through all of the digits on the calculator until the final step and then rounding to the proper number of significant figures. Failure to do this may cause your answer to be significantly different from those in the textbook and study guide.

5. When working with percentage, the first step is usually to convert the percent into a fraction. Don't forget the need for this conversion.

PRACTICE PROBLEMS

1. (L5) Make the following temperature conversions.

°F	°C	K
100.	_____	_____
_____	-33	_____
_____	_____	373

2. (**L2**) A lead block has a length of .22 meters, a width of 365 millimeters, a height of 4.8 decimeters, and a mass of 439 kilograms. Based on this information, what is the density of lead in grams/cubic centimeter?

3. (**L4**) If the density of alcohol is 0.789 g/mL, how many mL of alcohol must be measured out for an experiment that requires 2.50×10^2 kilograms of alcohol?

4. (**L3**) The fuel efficiency of a certain automobile is rated as 28.6 miles/gallon. Convert this value into kilometers/liter. (Hint: 1 kilometer = 0.62137 mile and 1 Liter = 1.056710 quarts)

5. (**L2**) The diameter of a single gold atom is 288 picometers. How many gold atoms would one have to set side by side in order to have a line of gold atoms one kilometer long?

6. (**L8**) Yellow brass consists of the elements copper and zinc. If a 555 kilogram sample of yellow brass is 67% copper, how many grams of zinc are present in this sample?

7. (**L4**) The annual production of sulfuric acid in the United States varies, but it is approximately 30,000,000 tons per year. If the density of concentrated sulfuric acid is 1.85 grams/mL, how many liters of concentrated sulfuric acid are produced each year? (Hint: 1 pound = 453.6 grams)

PRACTICE PROBLEM SOLUTIONS

1.

°F	°C	K
100.	37.8	311
-27	**-33**	240
212	1.0×10^2	**373**

2. First find the volume in cm^3.

$$\text{Volume} = \text{length} \times \text{width} \times \text{height}$$

Volume =
    0.22 m x 100 cm/m x 365 mm
            x 1 cm/10 mm x 4.8 dm x 10 cm/dm

$$V = 38544 \ cm^3$$

Now convert the mass to grams and substitute in the density equation.

$$\text{Density} = \text{mass/volume}$$

$$D = \frac{439 \ kg \times 1000 \ g/kg}{38544 \ cm^3}$$

$$D = 11 \ g/cm^3$$

3. This problem can be solved simply by substituting in the density equation.

$$D = M/V$$

rearrange the equation, insert the mass and density values given as well as the conversion factors for kg to g and mL to liters.

$$V = M/D = \frac{2.50 \times 10^2 \ kg \times 1000 \ g/kg}{0.789 \ g/mL}$$

$$V = 3.17 \times 10^5 \ mL$$

4. If you remember the conversion factors, this is a simple problem for dimensional analysis.

Mileage Rating

$$= \frac{28.6 \text{ miles}}{\text{gal}} \times \frac{1 \text{ km}}{0.62137 \text{ mile}} \times \frac{1 \text{ gal}}{4 \text{ qts}} \times \frac{1.056710 \text{ qt}}{1 \text{ Liter}}$$

Mileage Rating = 12.2 km/L

5. Dimensional analysis provides the key to this problem also.

Total length =
        number of gold atoms x length of one atom

Rearrange to isolate the unknown

number of gold atoms

$$= \frac{1 \text{ kilometer}}{\frac{288 \text{ pm}}{1 \text{ Au atom}} \times \frac{1 \text{ m}}{1 \times 10^{12} \text{ pm}} \times \frac{1 \text{ km}}{1000 \text{ m}}}$$

number of gold atoms = 3.47×10^{12} atoms

6. Since the brass consists of only two elements, the percentage of zinc can be found by subtracting

        percent zinc = 100 - percent copper
                     = 100 - 67
                     = 33%

Convert this percent to a fraction and multiply times the total mass

        mass of zinc = 0.33 x 555 kilograms

        mass of zinc = 180 kilograms

convert to grams
        mass of zinc = 180 kg x 1000 g/kg

        mass of zinc = 1.8×10^5 grams

7. Again, this problem is best done with dimensional analysis.

grams of H_2SO_4

$= 3 \times 10^7$ tons x 2000 lb/ton x 453.6 g/lb

grams of H_2SO_4 = 2.7216×10^{13} g

volume of H_2SO_4 = $\dfrac{2.7216 \times 10^{13} \text{ g}}{1.85 \text{ g/mL} \times 1000 \text{ mL/L}}$

volume of H_2SO_4 = 1×10^{10} liters

PRACTICE TEST

Put away all of your books and notes and work on this test as though it were an actual examination. Allow yourself 30 minutes to do the test. The answers are at the end of this chapter.

1. The temperature at the surface of the planet Venus is about 470°C. Convert this temperature to (a) degrees Fahrenheit and (b) kelvins.

2. Make the following conversions as indicated.

a. 45.0 mm = _____ cm b. 0.3 km = _____ cm

c. 3.9 kg of water has a volume of _____ liters.

3. Suppose that you wish to purchase a water bed that has the dimensions 2.45 m x 21.5 dm x 23 cm. How many kilograms of water does this bed contain?

4. The density of platinum metal is 21.5 g/cm^3. Calculate the mass of a block of platinum that has the dimensions 81.0 cm x 37 mm x 0.023 m.

5. The solubility of salt, NaCl, in water is 360 gram/liter. How many kilograms of salt will dissolve in 0.52 gallons of water?

6. Suppose that someone offers you a golden opportunity to invest in an industrial plant that

will extract gold from sea water. The only information available about the plant is that it is expected to process 100,000,000 kilograms of sea water a day. From a chemical handbook you learn that the density of sea water is 1.03 g/mL and sea water contains 4.0×10^{-5} mg/L of gold. What is the maximum number of grams of gold that the plant could possibly produce per day?

CONCEPT TEST ANSWERS

1. matter   2. atom   3. elements
4. solvent, solute   5. chemical
6. 1.00 kilograms
7. mass is extensive, the others are intensive.

8. 1.00×10^3 cubic centimeters
9. -273.15   10. 100 , 212
11. a. 1/1000   b. 1/100   c. 1000
    d. 1/10   e. 1×10^{-6} f. 1×10^{-12}

12. a. 2.05×10^3 mL   b. 1.2×10^2 m
    c. 445 mm   d. 0.450
    e. 22.5 L   f. 235 g
13. a. 2   b. 3   c. 1   d. 3
14. a. 2.7×10^2 b. 3.5×10^8
    c. 7.4×10^{-3} d. 5.5×10^{-4}
15. rusting and burning (a & c) are chemical changes; boiling and dissolving (b & d) are physical changes.

PRACTICE TEST ANSWERS

1. (**L5**) a. 880 $^{\circ}$F   b. 740 K
2. (**L3**) a. 4.50 cm   b. 3×10^4 cm
                  c. 3.9 liters
3. (**L2**) 1.2×10^3 kg   4. (**L2**) 1.5×10^4 grams
5. (**L2 & L3**) 0.71 kilograms of NaCl
6. (**L2**) 4 grams of gold per day

CHAPTER 2
ATOMS AND ELEMENTS

LEARNING GOALS:

1. Understand the basic assumptions of Dalton's Atomic Theory of Matter, including his hypothesis that is now known as the law of conservation of matter.

2. Know the names (including spelling) and symbols of the first 36 elements.

3. Given the atomic number, Z, and mass number, A, for an element having the symbol X, be able to represent that atom using the notation

$$^A_Z X$$

Also be familiar with the relationships involving atomic number, mass number, number of protons, number of neutrons, and number of electrons. For ions, be able to determine the number of electrons from the charge or vice versa, and also to represent the ion using a form like that shown above. (Sec. 2.2)

4. Be able to calculate the average atomic mass of an element from the relative abundances of the component isotopes and also calculate the isotopic abundances for elements where only two isotopes of known masses exist. (Secs. 2.3 & 2.4)

5. Be familiar with the periodic table and be able to locate the alkali metals, the alkaline earth metals, the transition metals, the halogens, and the rare gases on the table. Using the periodic table, be able to identify an element as a metal, a nonmetal, or a metalloid and also identify main group and transition elements. (Sec. 2.5)

6. Since grams, moles, and number of atoms (or molecules) are closely related, be able to convert from one of these to another for both elements and

compounds. In order to do this it will also be necessary to understand and use atomic weights and molecular (or formula) masses. (Sec. 2.6)

IMPORTANT NEW TERMS:

alkali metal (2.5)
alkaline earth metal (2.5)
atomic theory of matter
atomic mass unit (amu) (2.2)
atomic weight (2.4)
atomic number (2.2)
Avogadro's Number (2.6)
electron (2.2)
group (2.5)
halogen (2.5)
hydrated salts (2.11)
ionic compound (2.9)
isotopes (2.3)
law of conservation of matter
main group element (2.5)
mass number (2.2)
metal (2.5)
metalloid (2.5)
molar mass (2.6)
mole (2.6)
neutron (2.2)
nonmetal (2.5)
nucleus (2.2)
percent abundance (2.4)
percent composition (2.4)
period (2.5)
periodic table (2.5)
proton (2.2)
rare gas (2.5)
transition element (2.5)

CONCEPT TEST

1. Compounds are formed by the combination of different _____ in the ratio of small, whole numbers.

2. A(n) _____ is a form of matter that is composed of only one type of atom.

3. The law of conservation of matter states that
_____ can neither be created nor destroyed
in a chemical reaction.

4. The atomic masses of the elements are measured
relative to the mass of an atom of the element
_____ that has six protons and six
neutrons.

5. The _____ of a particular atom,
indicated by the symbol A, indicates the total
number of protons and neutrons in the atom.

6. Isotopes are atoms having the same _____
but different mass numbers.

7. _____ is the number of
particles in one mole of any substance.

8. A(n) _____ is the smallest unit of a
compound that retains the chemical characteristics
of the compound.

9. To find the molar mass of the compound Fe_2O_3, add
together the mass of _____ moles of iron atoms
and _____ moles of oxygen atoms.

10. No matter what substance is involved, one mole
always consists of the same number of _____ .

11. All atoms of the same element have the same
_____.

12. The elements in the periodic table are arranged
in order of increasing _____.

13. Write the correct name for the element
represented by each symbol:

a. Si _____   b. K _____

c. Mn _____   d. S _____

e. Na _____   f. Fe _____

14. Identify each element in the preceding question as a metal, a nonmetal, or a metalloid.

metals _____

nonmetals _____

metalloids _____

15. Which of the element names below is spelled incorrectly? (Note, there may be more than one.)

a. phosphorous

b. florine

c. beryllium

d. silicone

e. vanadium

f. clorine

STUDY HINTS:

1. Many students seem to have difficulty spelling the names of some specific elements. For example, the following names often cause problems (commonly confused letters are underlined): beryllium, fluorine, silicon, phosphorus, sulfur, chlorine, chromium, nickel, and zinc. Also, don't confuse the symbols of the elements having names that begin with S (sodium, sulfur, silicon, and scandium) or those that begin with P (potassium and phosphorus).

2. The authors of the textbook emphasize the importance of learning to convert from moles to grams and from grams to moles, but it is worth stressing this point again. This simple conversion is a basic step in a great many of the problems encountered in the early part of the course. By the end of the year this calculation will have been repeated many, many times. Learning to to do it well now will save a great deal of frustration during the coming semester.

3. Students sometimes become confused because a single chemical symbol can have several different meanings. For instance, Fe can mean

    a. one atom of iron,
    b. one mole of iron atoms, or
    c. one molar mass of iron.

When reading chemical symbols, remember that all of these interpretations are possible. You must decide which one is most appropriate in a given case.

4. The quantity that the textbook calls molar mass may also be referred to as gram molecular mass or gram molecular weight. If these terms are encountered, try to remember they are different names for basically the same quantity.

PRACTICE PROBLEMS

1. (L3) Determine the number of protons, neutrons, and electrons in the following species

   (a) $^{7}_{3}Li$ (b) $^{26}_{12}Mg^{2+}$

a. _3_ protons, _4_ neutrons, and _3_ electrons.

b. _12_ protons, _14_ neutrons, and _12_ electrons.

2. (L6) Complete the table.

	B	Ar	Cu
number of grams	1.9	4,510	0.021
atomic mass	10.811	39.948	63.546
moles	0.18	113	3.3×10^{-4}
number of atoms	1.1×10^{23}	6.80×10^{25}	2.0×10^{20}

3. (L2) What elements correspond to these symbols?

a. Ni _____   b. N _____

c. Cr _____   d. Li _____

4. (L5) Classify each of the elements in the above question as either a metal or a nonmetal.

metals _____   nonmetals _____

5. (L5) Classify each of the elements in the above question as either a typical or transition element.

typical _____   transition _____

16

6. (**L4**) The element neon has three stable isotopes, with masses and abundances as shown below. What is the average atomic weight of neon?

Exact Mass	Relative Abundance (%)
19.9924	90.02
20.9940	0.257
21.9914	8.82

7. (**L6**) Element 15 is found in a variety of useful products, including fertilizers, detergents, and even matches. In 1984 approximately 3.3×10^8 kilograms of this element was produced in the United States. What is this element and how many moles of it were produced?

8. (**L6**) Suppose someone spent one million dollars ($1,000,000) per second for one billion years (1,000,000,000 years), what fraction of Avogadro's number of dollars would have been spent?

PRACTICE PROBLEM SOLUTIONS

1. a. The atomic number is 3, and so the number of protons = 3. Since this is a neutral species, the number of electrons must also equal the atomic number, or 3.

The mass number, 7, must equal the sum of the number of protons and neutrons. Since the number of protons has already been found to be 3, then

mass number
       = number of protons + number of neutrons

   7   =   3   + number of neutrons

      number of neutrons = 7 - 3 = 4

1b. The atomic number is 12, so the number of protons = 12.

Since this is a charged species,
ion charge
      = number of protons - number of electrons
  +2 =   12   -   number of electrons

      number of electrons = 10

Finally, using the mass number relationship

mass number
      = number of protons + number of neutrons
  26 =   12   + number of neutrons

      number of neutrons = 14

2.   B   Ar   Cu

number
  of grams   1.9   4510   0.021

atomic mass   10.8   39.95   63.5

moles   0.18   113 3.3×10^{-4}

number
of atoms 1.1×10^{23} 6.80×10^{25} 2.0×10^{20}

18

3. a. nickel   b. nitrogen
   c. chromium   d. lithium
4. metals: Ni, Cr, Li   nonmetals: N
5. typical: N, Li   transition: Ni, Cr
6. atomic mass
   = (isotope 1 abundance)(isotope 1 mass)
     +(isotope 2 abundance)(isotope 2 mass)
     +(isotope 3 abundance)(isotope 3 mass)

atomic mass
   = 0.9002x19.9924 + 0.00257x20.9940
                    + 0.0882x21.9914

atomic mass = 19.99 grams/mole

7. From the periodic table, the element having an atomic number of 15 is phosphorus, and the molar mass of phosphorus is 30.9738 grams/mole.

$$\text{mol of P} = \frac{3.3x10^8 \text{ kg x 1000 g x 1 mole}}{1 \text{ kg x } 31.0 \text{ g}}$$

mol of P = $1.1x10^{10}$

8. First, calculate the total dollars spent

dollars spent =
        $1x10^6$ dollars/s x 60 s/min x 60 min/hr
             x 24 hr/day x 365 day/yr x $1x10^9$ yr

Total dollars = $3.2x10^{22}$ dollars

Now find what percent this is of $6.02x10^{23}$ dollars

$$\text{Percent} = \frac{3.2x10^{22} \text{ dollars x 100}}{6.02x10^{23} \text{ dollars}}$$

Percent = 5%

PRACTICE TEST
Allow 30 minutes to complete this practice test.

1. The element copper, which has an average atomic weight of 63.54 grams/mole, has two stable isotopes, copper-63 (mass = 62.9298) and copper-65

19

(mass = 64.9278).   What fraction of naturally occurring copper is Copper-63?

2. Determine how many protons, neutrons, and electrons are present in

a. $^{18}_{8}O^{2-}$ and b. $^{60}_{27}Co^{3+}$

3. Write the name of the element that matches each symbol and indicate whether each is a metal, metalloid, or nonmetal.

a. Zn   b. Sc   c. O   d. Al

4. Name each of the following elements and identify it as an alkali metal, an alkaline earth metal, a halogen, or a transition metal.

a. F   b. Li   c. Be   d. Ti   e. Ca

5. What is the mass in grams of a single atom of an element that has an atomic weight of 28.1 g/mole?

6. Complete the table.

number	Na	Si	Ne
of grams	_____	__67.2__	_____
atomic mass	_____	_____	_____
moles	_____	_____	__0.025__
number of atoms	5.5×10^{21}	_____	_____

CONCEPT TEST ANSWERS

1. atoms (or elements)   2. element
3. matter   4. carbon
5. mass number   6. atomic number

7. Avogadro's number(6.022×10^{23})
8. molecule   9. two, three
10. particles
11. atomic number (or number of protons)

20

12. atomic number (atomic mass is <u>not</u> acceptable.)
13. a. silicon   b. potassium
    c. manganese   d. sulfur
    e. sodium   f. iron
14. K, Mn, Na, and Fe are metals, S is a nonmetal,
    and Si is a metalloid.
15. phosphorus, fluorine, silicon, chlorine

PRACTICE TEST ANSWERS

1. **(L4)** 69.46% copper-63
2. **(L3)** a. 10 neutrons, 8 protons, and 10 electrons
            b. 33 neutrons, 27 protons, and 24 electrons
3. **(L2)** a. zinc   metal
            b. scandium   metal
            c. oxygen   nonmetal
            d. aluminum   metal
4. **(L5)** a. F   fluorine   halogen
            b. Li lithium   alkali metal
            c. Be beryllium   alkaline earth
            d. Ti titanium   transition metal
            e. Ca calcium   alkaline earth
5. **(L6)** 4.67×10^{-23} grams
6. **(L6)**

	Na	Si	Ne
number of grams	2.1×10^{-1}	**67.2**	5.1×10^{-1}
atomic mass	23.0	28.09	20.2
moles	9.1×10^{-3}	2.39	**0.025**
number of atoms	**5.5×10^{21}**	1.44×10^{24}	1.5×10^{22}

21

CHAPTER 3
COMPOUNDS AND MOLECULES

LEARNING GOALS:

1. Identify the elements most likely to exist as cations and those most likely to exist as anions. Be able to determine the charge for monatomic ions and predict the formulas for simple ionic compounds. Given a simple ionic compound, be able to recognize what component ions are present and give their relative number. (Sec. 3.3)

2. Know the names, formulas, and charge of the common polyatomic ions. (Sec. 3.3)

3. Know how to name ionic compounds and binary compounds of the nonmetals using the rules explained in the textbook. (Sec. 3.4)

4. Understand the differences between a molecular formulas and an empirical formula, between molar mass and formula mass. Given the formula of a compound you should be able to determine the molar mass (or formula mass). (Secs. 3.5 & 3.6)

5. The relationships that allow conversions among quantities such as grams, moles, number of molecules, and number of atoms of component elements are among the most fundamental in chemistry. Be sure that you understand these relationships. (Sec. 3.5)

6. The law of constant composition allows us to define a compound in terms of either its percent composition or its empirical formula. Be able to calculate percent composition from empirical formula or vice versa. You should also understand the common methods for calculating the molecular formula from the empirical formula. (Sec. 3.6)

7. You should be able to apply the concepts of molecular composition outlined in goal 6 to special cases, such as the composition of hydrated compounds. (Sec. 3.6)

Important New Terms:

anhydrous compounds (3.6)
anions (3.3)
binary compounds (3.4)
cations (3.3)
Coulomb's law (3.3)
diatomic molecule (3.2)
empirical formula (3.6)
formula weight (3.5)
hydrated compounds (3.6)
ionic compound (3.3)
ions (3.3)
law of multiple proportions
law of constant composition (3.6)
molecular formula (3.6)
molecular weight (3.4)
monatomic species (3.3)
percent composition (3.6)
polyatomic species (3.1)

CONCEPT TEST

1. Circle the symbol or symbols below for the pure elements that are <u>not</u> diatomic.

He   H   I   P   O   Na

2. Circle the hydrated compound or compounds in the following list.

a. H_2O b. $NiCl_2 \cdot 6H_2O$ c. $HClO_4$

     d. $BaCl_2 \cdot 2H_2O$ e. $Ca(OH)_2$

3. Metal atoms usually lose _____ to form positively charged ions called _____.

4. How many moles of oxygen atoms are present in one mole of each of the compounds below?

a. H_2O b. $NiCl_2 \cdot 6H_2O$ c. $HClO_4$

     d. $BaCl_2 \cdot 2H_2O$ e. $Ca(OH)_2$

5. Which comes first when writing a chemical formula, the symbol of the cation or the symbol of the anion? _____

6. _____ law describes the attractive force between oppositely charged ions.

7. Nonmetals give ionic compounds only when combined with a(n) _____.

8. The _____ is the mass in grams of Avogadro's number of molecules.

9. The law of constant composition states that any sample of a pure compound always consists of the same _____ combined in the same _____ _____.

10. The molecular formula is always the product of _____ multiplied times the subscripts in the empirical formula.

11. To find the molar mass of the compound Fe_2O_3, you must add together the mass of _____ moles of iron atoms and _____ moles of oxygen atoms.

12. A(n) _____ is the smallest unit of a compound that retains the chemical characteristics of the compound.

13. The sum of the atomic weights of all of the atoms in a molecule is called the _____ _____ and is numerically equivalent to the molar mass of the compound.

14. Name each of these polyatomic ions:

a. NO_3^- _____   b. OH^- _____

c. SO_4^{2-} _____   d. PO_4^{3-} _____

15. Complete the table below by writing the formula for each polyatomic ion:

a. ammonium _____   b. sulfate _____

c. acetate _____   d. carbonate _____

STUDY HINTS:

1. Remember that under normal conditions pure hydrogen, nitrogen, oxygen, and the halogens exist as diatomic molecules, whereas the rare gas elements exist as uncombined atoms.

2. When working problems related to chemical formulas, always remember that a formula is basically a mole ratio. Therefore, the initial stage of such a problem often involves the determination of the number of moles of the components in the formula.

3. Like chemical symbols, chemical formulas can have several different meanings. For instance, H_2O can mean

    a. one molecule of water,
    b. two atoms of the element hydrogen combined with one atom of oxygen,
    c. one mole of water molecules, or
    d. one molar mass of water molecules.

Be sure that you consider these possibilities when you read either a chemical symbol or formula.

4. Students often have difficulty with problems involving a mass decrease due to the formation of a gas, such as oxygen or carbon dioxide. Once you learn to recognize this problem type, it really isn't as difficult as it looks. Just remember that the decrease in mass is due to the escape of a gas, so the mass change represents the mass of gas produced. Use this value to find the moles of gas, which is usually the key to solving the problem. Also be alert for a variation of this problem type where the mass increases due to the combination of the original solid with a gas, like oxygen. Again, the key step is to calculate the moles of gas from the mass change.

PRACTICE PROBLEMS

1. (L4) How many moles of phosphorus atoms are contained in one mole of each of the following compounds?

a. P_2O_5 _2_   b. $Ca_3(PO_4)_2$ _2_   c. $Na_5P_3O_{10}$ _3_

2. (**L2 & L3**) Name each compound

a. MnS_2 _Manganese (II) Sulfide_ b. $KMnO_4$ _Pottassium Permanganate_

c. $TiCl_4$ _Titanium (IV) Chloride_ d. NaCN _Sodium Cynide_

3. (**L2 & L3**) Write the formula for each compound

a. ammonium nitrate _NH_4NO_3_

b. iron(III) chloride _$FeCl_3$_

c. oxygen difluoride _OF_2_

d. vanadium(IV) sulfate _$V(SO_4)_2$_

e. aluminum trichloride _$AlCl_3$_

4. (**L1**) Write formulas for all of the compounds that can be formed by combining each cation with each of the anions listed. Name the resulting compounds.

CATIONS

Li^+
Ca^{2+}

Answers:

_____

_____

ANIONS

F^-
PO_4^{3-}

_____

_____

5. (**L5**) Assume that you have 15 grams of each of the following compounds. How many moles does that represent in each case?

a. NH_3 0.88 moles

b. NH_4NO_3

c. NH_2COONH_4

6. (**L5**) Hydrazine, N_2H_4, was used as a rocket fuel by the Germans in World War II.

a. What is the molar mass of hydrazine? _____

26

b. How many moles of hydrazine
are contained in a pure sample
of N_2H_4 having a mass of 9.6 grams? _____
c. How many molecules of N_2H_4
are present in this 9.6 gram sample? _____
d. How many hydrogen atoms are
present in a 9.6 gram sample
of N_2H_4? _____

7. (L6) The empirical formula of a certain compound
is CH_2 and the molar mass is 84.0 g/mole. What is
the molecular formula of this compound?

8. (L6) The first true compound of a rare gas was
isolated in 1962 and found to have the composition
29.8% xenon, 44.3% platinum, and 25.9% fluorine.
What is the empirical formula of this compound?

9. (L7) Natural gypsum is a hydrated form of
calcium sulfate having the formula $CaSO_4 \cdot xH_2O$. When
2.00 grams of gypsum is heated at $200^\circ C$ until all of
the water has been driven off, the mass of the
remaining solid is 1.58 grams. What is the value
of x in the formula for gypsum?

PRACTICE PROBLEM SOLUTIONS

1. a. two moles   b. two moles   c. three moles

2. a. manganese(IV) sulfide
   b. potassium permanganate
   c. titanium(IV) chloride
   d. sodium cyanide

3. a. NH_4NO_3 b. $FeCl_3$ c. OF_2
   d. $V(SO_4)_2$ e. $AlCl_3$

4.
   LiF   lithium fluoride
   Li_3PO_4 lithium phosphate
   CaF_2 calcium fluoride
   $Ca_3(PO_4)_2$ calcium phosphate

5. a. NH_3 15 g/17.0 g/mol = 0.88 moles

   b. NH_4NO_3 15 g/79.0 g/mol = 0.19 moles

   c. NH_2COONH_4 15 g/78.0 g/mol = 0.19 moles

6. a. 32 g/mole   b. 3.0×10^{-1} mole
   c. 1.8×10^{23} molecules   d. 7.2×10^{23} atoms

7. The apparent molar mass of CH_2 is 14
To find how many times this unit is contained in
84 g/mole, divide

$$\frac{84}{14} = 6$$

Therefore the molecular formula is 6 times the
empirical formula, or

   C_6H_{12}

8. First assume 100.0 grams of compound in order to
obtain masses of the component elements

mole Xe = 29.8 g / 131.3g/mole = 0.2270 mole Xe

mole Pt = 44.3 g / 195.1 g/mole = 0.2271 mole Pt

mole F = 25.9 g / 19.0 g/mole = 1.363   mole F

To obtain the mole ratio, divide each number of
moles by 0.2270, the smallest number of moles.

28

Xe $\dfrac{0.2270}{0.2270} = 1$ Pt $\dfrac{0.2271}{0.2270} = 1$ F $\dfrac{1.363}{0.2270} = 6$

Thus the ratio Xe:Pt:F :: 1:1:6,
and the formula is

$$XePtF_6$$

9. To find x, it is necessary to calculate the ratio of the moles of water to the moles of nonhydrated $CaSO_4$.

$$\text{mole } H_2O = \dfrac{2.00g - 1.58 g}{18.0 \text{ g/mole}} = 0.02333 \text{ moles } H_2O$$

$$\text{mole } CaSO_4 = \dfrac{1.58 g}{136 \text{ g/mole}} = 0.01160 \text{ mole } CaSO_4$$

$$\dfrac{\text{mole water}}{\text{mole calcium sulfate}} = \dfrac{0.02333 \text{ moles}}{0.01160 \text{ moles}} = 2$$

$\underline{x = 2}$ and the formula is $CaSO_4 \cdot 2H_2O$

PRACTICE TEST
(Allow yourself 35 minutes)

1. How many moles of sulfur atoms are contained in one mole of each of the following compounds?

a. CS_2 b. $Na_2S_2O_4$ c. $Ca(HSO_3)_2$

2. Sodium thiosulfate, which has the formula $Na_2S_2O_3 \cdot 5H_2O$, is used in the photographic and paper-making industries. What is the molar mass of sodium thiosulfate and the percentage of sulfur in this compound.

3. Write the formula for each compound listed.

a. boron trichloride   b. copper(II) chloride
c. aluminum nitrate   d. sodium sulfide

4. Write the name for each compound listed.

a. KCH_3COO b. $TiCl_4$
c. CaI_2 d. NH_4ClO_4

5. For each ionic compound, give the formula, charge, and number of each component ion.
a. $(NH_4)_2SO_3$ b. $CaCl_2$ c. K_2SO_4 d. $Mg_3(PO_4)_2$

6. Answer the following questions regarding the compound phosgene, $COCl_2$, a World War I poison gas.

a. What is the molar mass of
phosgene.   _____
b. How many moles of phosgene
are contained in a pure sample
of $COCl_2$ having a mass of 2.00 grams?   _____
c. How many molecules of $COCl_2$
are present in this 2.00 gram sample?   _____
d. How many chlorine atoms are
present in a 2.00 gram sample
of $COCl_2$?   _____

7. Fertilizers are one of the chemical products produced in the greatest quantity. Determine which of the two phosphorus-containing compounds listed will contain the higher percentage of phosphorus.
a. $Ca(H_2PO_4)_2$ b. $NH_4H_2PO_4$

8. Calculate the atomic weight of the unknown element M, if the molar mass of the compound $Na_2M_2O_3$ is 156 g/mol.

9. Mirex is a pesticide that breaks down slowly in the environment but is very useful for controlling certain insect pests. Mirex is composed of 21.9% carbon and 78.1% chlorine, and has a molar mass of 545 g/mol. (a) what is the empirical formula of mirex? (b) What is the molecular formula of mirex?

10. When solid magnesium carbonate is heated, a portion of it decomposes to form solid magnesium oxide and gaseous carbon dioxide. If 2.25 grams of magnesium carbonate is heated until the remaining solid, a mixture of MgO and unreacted $MgCO_3$, has a mass of 1.95 grams, what percentage of the $MgCO_3$ has decomposed?

CONCEPT TEST ANSWERS

1. Helium (He), phosphorus (P), and Sodium (Na) are not diatomic.

2. $NiCl_2 \cdot 6H_2O$ and $BaCl_2 \cdot 2H_2O$ are hydrates.

3. electrons, cations
4. a. one b. six c. four d. two e. two
5. the cation   6. Coulomb's
7. metal   8. molar mass
9. elements, proportions by mass
10. a small, whole number   11. two, three
12. molecule   13. molecular weight
14. a. nitrate   b. hydroxide
    c. sulfate   d. phosphate
15. a. NH_4^+ b. SO_4^{2-}

    c. $C_2H_3O_2^-$ d. CO_3^{2-}

PRACTICE TEST ANSWERS

1. **(L5)** Each compound contains two moles of sulfur per mole of compound.
2. **(L4 & L6)** molar mass = 248.17 g/mol
   The percent of sulfur is 25.84%
3. **(L2 & L3)** a. BCl_3 b. $CuCl_2$

   c. $Al(NO_3)_3$ d. Na_2S

4. **(L2 & L3)** a. potassium acetate
   b. titanium(IV) chloride
   c. calcium iodide
   d. ammonium perchlorate
5. **(L1)** a. $(NH_4)_2SO_3$: two NH_4^+ and one SO_3^{2-}

   b. $CaCl_2$: one Ca^{2+} and two Cl^-

   c. K_2SO_4 : two K^+ and one SO_4^{2-}

   d. $Mg_3(PO_4)_2$: three Mg^{2+} and two PO_4^{3-}

6. **(L5)** a. 98.91 g/mol   b. 0.0202 mole

   c. 1.22×10^{22} molecules d. 2.44×10^{22} atoms
7. **(L6)**   % phosphorus
   a. $Ca(H_2PO_4)_2$ 26.47%
   b. $NH_4H_2PO_4$ 26.92%

The second compound is slightly higher.

8. **(L4)** 31 g/mol
9. **(L6)** a. C_5Cl_6 b. $C_{10}Cl_{12}$
10. **(L7)** percent decomposed = 25.5%

CHAPTER 4
CHEMICAL REACTIONS: AN INTRODUCTION

LEARNING GOALS:

1. Know how to balance simple chemical equations.
(Sec. 4.2)

2. Be able to predict the products and balance the
resulting equation for some common reactions, such
as the combination of a metal with oxygen or a
halogen, the combustion of a hydrocarbon, and the
decomposition of a metal carbonate due to heating.
(Sec. 4.3)

3. Use a balanced chemical equation to write the
stoichiometric factors for a chemical reaction and
use these values to calculate the relationships
between the moles or mass of products and
reactants. (Sec. 4.4)

4. In some stoichiometric calculations the amount
of one of the reactants causes that reagent to be
the controlling factor in determining the extent to
which the reaction will occur. This is called a
limiting reagent problem. It's necessary to be
able to recognize this type of problem, to identify
the limiting reagent, and to use the amount of
limiting reagent to perform the calculations
described in learning goal 3. (Sec. 4.5)

5. Understand what is meant by actual yield,
theoretical yield, and percent yield, and use the
equation

$$\text{percent yield} = \frac{\text{theoretical yield x 100}}{\text{actual yield}}$$

to do simple calculations. (Sec. 4.6)

6. Some applications of the stoichiometric
principles discussed in the chapter are the use of
combustion analysis to determine an empirical
formula or the composition of a mixture. Know how
to work this type of problem. (Sec. 4.7)

IMPORTANT NEW TERMS:

actual yield (4.6)
balanced chemical equation (4.1)
chemical equation (4.1)
combination reaction (4.3)
combustion analysis (4.7)
combustion reaction (4.3)
decomposition reaction (4.3)
limiting reagent (4.5)
percent yield (4.6)
product (4.1)
reactant (4.1)
stoichiometric coefficient (4.4)
stoichiometry (4.4)
theoretical yield (4.6)

CONCEPT TEST

1. A(n) _____ is a combination of chemical symbols and formulas used to represent a chemical reaction.

2. According to the principle of _____, the total mass of the products in a chemical reaction must equal the total mass of the original reactants.

3. The products of the <u>complete</u> combustion of hydrocarbons are always the compounds _____ and _____ .

4. Chemical reactions in which compounds break down into simpler compounds are called _____ _____. Often this occurs due to heating.

5. When a metal carbonate reacts with acid, the products are a(n) _____ and _____.

6. _____ is the name given to the mole-ratio factor relating moles of a desired compound to moles of an available reagent.

7. Combustion reactions involve the combination of some reactant (or reactants) with _____ (name the element).

8. Whenever possible, what is the first step in any quantitative chemical calculation? _____ _____

9. When one reactant in a chemical reaction is present in an amount less than that required by the stoichiometry of the equation, that reactant is called the _____ and it will determine the quantity of products formed.

10. The maximum amount of product that can be isolated from a given quantity of reactants is called the _____.

11. The efficiency of a chemical reaction is often evaluated in terms of the _____ , that is, the actual yield times 100 divided by the theoretical yield.

12. If the initial moles of methane (natural gas) reacting is known, to calculate how many moles of oxygen are required to react with it in the balanced equation

$$CH_4 + 2 O_2 \rightarrow CO_2 + 2 H_2O$$

the stoichiometric factor is _____.

STUDY HINTS:

1. Whenever you begin to work on a stoichiometry problem, your first step should be to make sure that any equations provided are complete and balanced. In problems like this, an equation that hasn't been balanced is a mistake waiting to happen.

2. It is very important to realize that the mole

34

ratio in the balanced equation is the key to doing most of the problems in this chapter. As the textbook points out, most of these problems can be thought of in three steps: first, convert whatever you are given into moles; second, use the stoichiometric ratio to convert to moles of another substance in the balanced equation; and third, convert from moles of that new substance to whatever units are requested.

3. Notice that the limiting reagent is not necessarily the reactant present in the smallest amount; you must also consider the stoichiometric factor. The reagent present in the largest amount may be the limiting reagent if there is not enough of it to satisfy the requirements of the balanced equation. Watch for a practice problem that illustrates this.

PRACTICE PROBLEMS

1. (L1) Balance each equation.

a. _1_ $Sn(s)$ + _2_ $Cl_2(g)$ → _1_ $SnCl_4(s)$

b. _4_ $P(s)$ + _5_ $O_2(g)$ → _1_ $P_4O_{10}(s)$

c. _2_ $NH_3(g)$ + ___ $O_2(g)$ → _2_ $NO(g)$ + _3_ $H_2O(g)$

2. (L2) Complete the following equations by writing the formula of the product (or products) and balancing.

a. _2_ $\overset{2+}{Ca}(g)$ + _2_ $\overset{2-}{O_2}(g)$ → _2_ CaO _____

b. _2_ $\overset{1+}{K}(s)$ + _1_ $\overset{1-}{Cl_2}(g)$ → KCl _____

c. ___ $C_2H_4(g)$ + ___ $O_2(g)$ → H_2O + CO_2

d. ___ $LiCO_3(s)$ + heat → LiO + CO_2

3. (L3) Answer the following question based on the balanced equation

$Al_4C_3(s)$ + 12 $H_2O(\ell)$ → 4 $Al(OH)_3(aq)$ + 3 $CH_4(g)$

6.42mol.   265g   ?   128g
            ?mol.

35

a. How many moles of water are necessary to react completely with 6.42 moles of Al_4C_3?

b. How many moles of $Al(OH)_3$ will result from the complete reaction of 265 grams of water with excess aluminum carbide, Al_4C_3?

c. How many grams of H_2O are required to prepare 128 grams of CH_4 using the above reaction?

4. (L4) Crude Titanium metal is prepared commercially according to the balanced equation

$$TiCl_4 \;+\; 2\ Mg \;\rightarrow\; 2\ MgCl_2 \;+\; Ti$$

8.52×10⁴g   4.0×10⁴g

If 40.0 kilograms of magnesium is reacted with 85.2 kilograms of titanium chloride, (a) what is the limiting reagent, and (b) how many grams of titanium metal will be formed?

5. (**L5**) In the previous problem, the theoretical yield of titanium metal is 21.5 kilograms. If the actual yield of titanium metal in the experiment was 10.4 kilograms, what was the percent yield for this reaction?

6. (**L6**) A certain compound that consists only of carbon and hydrogen is burned in air, producing 8.80 grams of carbon dioxide and 7.20 grams of water. What is the empirical formula of this compound?

PRACTICE PROBLEM SOLUTIONS

1. a. $Sn(s)$ + 2 $Cl_2(g) \rightarrow SnCl_4(s)$

   b. 4 $P(s)$ + 5 $O_2(g) \rightarrow P_4O_{10}(s)$

   c. 4 $NH_3(g)$+ 5 $O_2(g) \rightarrow$ 4 $NO(g)$ + 6 $H_2O(g)$

2. a. 2 $Ca(g)$ + $O_2(g)$ $\rightarrow$ 2 $CaO(s)$

   b. 2 $K(s)$ + $Cl_2(g)$ $\rightarrow$ 2 $KCl(s)$

   c. $C_2H_4(g)$ + 3 $O_2(g) \rightarrow$ 2 $CO_2(g)$ + 2 $H_2O(g)$

   d. $LiCO_3(s)$+ heat $\rightarrow$ $LiO(s)$ + $CO_2(g)$

37

3. a. According to the balanced equation, 12 moles of water is necessary to react completely with 1 mole of Al_4C_3.

moles of water = 6.42 mol Al_4C_3 x $\dfrac{12 \text{ mol water}}{1 \text{ mol } Al_4C_3}$

__moles water = 77.0 moles__

b. First, determine the number of moles of water.

moles of water = $\dfrac{265 \text{ g}}{18.02 \text{ g/mol}}$

moles of water = 14.71 mol

From the equation 12 mol of water produces 4 mol of Al_4C_3, so using this stoichiometric factor

mol of Al_4C_3
    = 14.71 mol water x $\dfrac{4 \text{ mol aluminum carbide}}{12 \text{ mol water}}$

moles of Al_4C_3 = __4.90 moles__

c. As before, the first step is to determine the moles of CH_4

moles of CH_4 = $\dfrac{128 \text{ g}}{16.04 \text{ g/mol}}$ = 7.975 moles

From the balanced equation, 12 moles of water are required to produce 3 moles of methane.

moles of H_2O = 7.975 moles CH_4 x $\dfrac{12 \text{ mol water}}{3 \text{ mol } CH_4}$

moles of H_2O = 31.90 mol

Finally, determine the grams of water

grams of water = 31.90 moles x 18.02 g/mol

__grams of water = 575 g__

4. a. The first step in determining the limiting reagent is to calculate the number of moles of each reagent available.

$$\text{moles } TiCl_4 = \frac{85.2 \text{ kg} \times 1000 \text{ g/kg}}{189.7 \text{ g/mol}} = 449.1 \text{ moles}$$

$$\text{moles Mg} = \frac{40.0 \text{ kg} \times 1000 \text{ g/kg}}{24.31 \text{ g/mol}} = 1645 \text{ moles}$$

Using the balanced equation, calculate the ratio of moles required.

$$\text{ratio of moles required} = \frac{2 \text{ moles Mg}}{1 \text{ mol } TiCl_4} = 2.0$$

Compare this with the moles available

$$\text{ratio of moles available} = \frac{1645 \text{ moles Mg}}{449.1 \text{ moles } TiCl_4} = 3.66$$

The "ratio of moles required" is smaller than the "ratio of moles available" so the reagent in the denominator is the limiting reagent.
   __Titanium(IV) Chloride is the limiting reagent.__

b. The balanced equation indicates that one mole of $TiCl_4$ will form one mole of Ti, so the next step is easy.

moles Ti formed = moles of $TiCl_4$ = 449.1 moles

Now, convert the moles of Ti to grams of Ti.

mass of Ti = 449.1 moles of Ti x 47.88 g/mol

__mass of Ti = 2.15×10^4 g__

5. Remember that the definition of percent yield is

$$\text{percent yield} = \frac{\text{actual yield} \times 100}{\text{theoretical yield}}$$

Substituting into this equation

$$\text{percent yield} = \frac{10.4 \text{ kg} \times 100}{21.5 \text{ kg}}$$

percent yield = 48.4 %

6. All of the carbon in the original compound has been converted into carbon dioxide, and all of the hydrogen into water. Therefore, it's apparent that

moles of carbon = moles of carbon dioxide formed

moles of hydrogen = 2 x moles of water formed

Next calculate the moles of water and carbon dioxide.

$$\text{mol } CO_2 = \frac{8.80 \text{ g}}{44.01 \text{ g/mol}} = 0.2000 \text{ moles}$$

$$\text{mol } H_2O = \frac{7.20 \text{ g}}{18.02 \text{ g/mol}} = 0.4000 \text{ moles}$$

Determine the moles of hydrogen and carbon

$$\text{mol } C = \frac{1 \text{ mol C}}{1 \text{ mol } CO_2} \times 0.2000 \text{ mol } CO_2$$

$$= 0.2000 \text{ mol of C}$$

$$\text{mol } H = \frac{1 \text{ mol H}}{2 \text{ mol } H_2O} \times 0.4000 \text{ mol } H_2O$$

$$= 0.8000 \text{ mol of H}$$

Divide the number of moles of each element by the smallest number of moles, that is, 0.200

$$\frac{0.8000 \text{ mol H}}{0.2000 \text{ mol C}} = \frac{4 \text{ mol H}}{1 \text{ mol C}}$$

The empirical formula is CH_4

PRACTICE TEST (40 Minutes)

1. Balance the following equations that describe chemical processes used in industry.
a. formation of dilute phosphoric acid

_____ $P_4O_{10}(s)$ + _____ $H_2O(\ell)$ → _____ $H_3PO_4(aq)$

b. catalyzed reduction of nitrogen oxide emissions

___ $NO(g)$ + ___ $NO_2(g)$ + ___ $NH_3(g)$
→ ___ $N_2(g)$ + ___ $H_2O(g)$

c. production of white phosphorus from apatite

_____ $Ca_3(PO_4)_2(s)$ + _____ $C(s)$
→ ___ $CaO(s)$ + ___ $CO(g)$ + ___ $P(s)$

2. Complete and balance the equations for each of the following reactions.

a. $Mg(s)$ + $O_2(g)$ →   _____

b. $BaCO_3(s)$ + heat   →   _____ + _____

c. $Li(s)$ + $F_2(g)$ →   _____

d. $C_4H_{10}(g)$ + $O_2(g)$ →   _____ + _____

3. Pure silicon is produced for transistors by reacting $SiCl_4$ with magnesium according to the equation

$SiCl_4(g)$ + 2 $Mg(s)$ → 2 $MgCl_2(s)$ + $Si(s)$

How many moles of magnesium are required to react completely with 34 grams of $SiCl_4$?

4. Nitric acid can be produced by reacting nitrogen dioxide with water according to the equation

3 $NO_2(g)$ + $H_2O(\ell)$ → 2 $HNO_3(aq)$ + $NO(g)$

How many grams of HNO_3 can be produced by reacting 13.8 grams of NO_2 with an excess of water?

5. Fibers of boron carbide are used in some bullet-proof protective clothing. Boron carbide can be synthesized from boron oxide by means of the following two step process

$$B_2O_3(s) + 3\ C(s) + 3\ Cl_2(g) \rightarrow 6\ CO(g) + 2\ BCl_3(g)$$

$$4\ BCl_3(g) + 6\ H_2(g) + C(s) \rightarrow B_4C(s) + 12\ HCl(g)$$

How many moles of B_4C could be produced from the complete reaction of 250 moles of B_2O_3 according to the sequence of reactions given above?

6. Iron can be obtained in a blast furnace by the reaction of the ore hematite with carbon in a series of reactions that may be represented by the overall equation

$$2\ Fe_2O_3 + 3\ C \rightarrow 3\ CO_2 + 4\ Fe$$

If a company has a stockpile of 3.27×10^6 kg of hematite and 5.85×10^5 kg of carbon, how many grams of iron can be produced?

7. Xenon and fluorine are reacted to form 37.9 grams of xenon tetrafluoride. Based on the masses of the starting materials, the expected yield of xenon tetrachloride was 64.5 grams. What is the percent yield of this process?

8. A certain compound that consists only of carbon and hydrogen is burned in air, producing 13.2 grams of carbon dioxide and 7.20 grams of water. What is the empirical formula of this compound?

CONCEPT TEST ANSWERS

1. chemical equation
2. conservation of mass
3. carbon dioxide and water
4. Decomposition reactions
5. salt and carbonic acid
   (Notice the carbonic acid forms water and CO_2)
6. stoichiometric factor
7. oxygen
8. balance any equations involved

9. limiting reagent
10. theoretical yield
11. percent yield
12. two

PRACTICE TEST ANSWERS

1. (**L1**) a. $P_4O_{10}(s) + 6 H_2O(\ell) \rightarrow 4 H_3PO_4(\ell)$

   b. $NO(g) + NO_2(g) + 2 NH_3(g) \rightarrow 2 N_2(g) + 3 H_2O(g)$

   c. $Ca_3(PO_4)_2(s) + 5 C(s)$
   $\rightarrow 3 CaO(s) + 5 CO(g) + 2 P(s)$

2. (**L2**) a. $2 Mg(s) + O_2(g) \rightarrow 2 MgO(s)$

   b. $BaCO_3(s) + heat \rightarrow BaO(s) + CO_2(g)$

   c. $2 Li(s) + F_2(g) \rightarrow 2 LiF(s)$

   d. $2 C_4H_{10}(g) + 13 O_2(g) \rightarrow 8 CO_2(g) + 10 H_2O(\ell)$

3. (**L3**) 0.40 moles   4. (**L3**) 12.6 g
5. (**L3**) 125 moles   6. (**L4**) 2.29×10^9 grams
7. (**L5**) 58.8%   8. (**L6**) C_3H_8

CHAPTER 5
REACTIONS IN AQUEOUS SOLUTION

LEARNING GOALS:

1. Know the basic terms used to describe solutions, including terms encountered before, such as solute, solvent and solution, as well as new terms, like strong and weak electrolytes, acids, and bases. Be able to use the solubility guidelines to predict the likelihood that an ionic compound will be soluble in water. (Sec. 5.1)

2. Memorize the names and formulas of the common acids and bases, as given in Table 5.1, including remembering whether each is a weak or strong electrolyte. (Sec. 5.1)

3. Be able to determine which ions will form when an ionic compound is dissolved in water and to write net ionic equations for chemical reactions involving ionic compounds. (Sec. 5.2)

4. Identify the important types of reactions in aqueous solution, including exchange reactions, precipitation reactions, acid-base reactions, and gas-forming reactions. In many cases it should also be possible to predict the products of these equations. (Sec. 5.3)

5. Be able to write equations for the preparation of compounds using exchange reactions, precipitation reactions, or acid-base reactions. (Sec. 5.3)

6. Solution concentrations are frequently expressed in terms of molarity, where

$$\text{molarity (M)} = \frac{\text{moles of solute}}{\text{volume of solution (L)}}$$

Use this equation to do simple solution calculations, including those required to prepare a solution of a given concentration either by dissolving the solute directly in a solvent or by dilution of a previously known solution. (Sec. 5.4)

7. Using a balanced equation and the molarity of the reactants, be able to perform solution stoichiometry calculations for titration and precipitation reactions. (Sec. 5.4)

8. Be able to determine the oxidation numbers of the elements in compounds. If given an oxidation-reduction reaction, be able to identify the oxidized and reduced substances, balance the equation in either acidic or basic solution, and do stoichiometric calculations. (Sec. 5.5)

IMPORTANT NEW TERMS:

acid (5.1)
acid-base reaction (5.3)
acid-base indicator (5.4)
base (5.1)
charge balance (5.5)
dilution (5.4)
dissociation (5.1)
electrolyte, strong (5.1)
electrolyte, weak (5.1)
endpoint (5.4)
equivalence point (5.4)
exchange (or displacement) reaction (5.3)
gas-forming reaction (5.3)
hydronium ion (5.1)
hydroxide ion (5.1)
mass balance (5.5)
molarity (5.4)
net ionic equation (5.2)
nonelectrolyte (5.1)
oxidation number (5.5)
oxidation-reduction reaction (5.5)
oxidized substance (5.5)
oxidizing agent (5.5)
precipitation reaction (5.3)
qualitative chemical analysis (5.4)
quantitative chemical analysis (5.4)
reduced substance (5.5)
reducing agent (5.5)
solubility guidelines (5.1)
solution stoichiometry (5.4)

spectator ion (5.2)
standardization (5.4)
titration (5.4)

CONCEPT TEST

1. A(n) _____ is a compound that forms an aqueous solution that doesn't conduct an electric current.

2. Ionic compounds that dissociate completely into ions in water and are good conductors of electricity are called _____ electrolytes.

3. Ions that appear in <u>exactly</u> the same form on the reactant and product sides of an ionic equation are called _____.

4. _____ is the name we give to the class of compounds that produces hydroxide ions when dissolved in water.

5. Write the general net ionic equation for the neutralization reaction of any strong acid with any strong base. _____

6. Dilution problems are based on the idea that the number of _____ of solute in the initial solution must equal the number of _____ of solute in the final solution.

7. That point in an acid-base titration at which the number of moles of OH^- added is equal to the number of moles of H_3O^+ present is called the

_____.

8. The endpoint of an acid-base titration is often detected by adding a dye called a(n) _____ that changes color at a hydronium ion concentration as close as possible to the equivalence point.

9. _____ is name given to the procedure used to determine the exact concentration of an acid, base, or other reagent, usually in preparation for using that substance in a titration.

10. Oxidation-reduction reactions are those involving the transfer of _____.

11. When a substance accepts electrons during a chemical reaction, it is said to be _____. The substance oxidized in a chemical reaction is called the _____ agent.

12. _____ is defined as the charge an atom has, or appears to have, when the electrons of the compound are counted according to a certain set of rules.

13. For each ion or compound determine the oxidation number for each element. (Don't forget the sign!)

a. $TiCl_4$ _____   b. SO_4^{2-} _____

c. $HClO_4$ _____   d. SiH_4 _____

14. Identify the oxidized element and the reduced element in each equation.

a. H_2O + Cl_2 → HCl + O_2
element oxidized _____   element reduced _____

b. Sn + HNO_3 → $Sn(NO_3)_2$ + NO + H_2O
element oxidized _____   element reduced _____

15. Identify the oxidizing agent and the reducing agent for each equation in the previous question.

<u>oxidizing agent</u>   <u>reducing agent</u>

a. _____   _____

b. _____   _____

STUDY HINTS:

1. Students will often confuse the weak electrolyte ammonia, NH_3, with the ammonium ion, NH_4^+.

2. Learning goal 7 suggests that you should be able to predict the products of acid-base exchange reactions. It is also helpful to be able to work backwards, predicting what acid and base would have to be reacted in order to form a given salt. Practice problem 5 tests your ability to do this. Later on in the course this skill will be very useful.

3. When you learn the definition of molarity, be sure to learn moles <u>of solute</u> per liter <u>of solution</u>, not just moles per liter. Remember that when preparing solution, liters of solvent added is not always the same as the final volume of the solution. In addition, you will soon learn another concentration unit with a name and definition similar to molarity. If you learn the complete definition for molarity now, you are less likely to be confused at that time.

4. Dilution problems should be fairly easy once you memorize the formula, but there is one frequent point of confusion. You must read carefully to distinguish between the volume of water added and the volume of the final solution. The three sections of practice problem 9 show how this type of question can be asked several different ways.

5. Remember that the oxidized substance is the reducing agent, and the reduced substance is the oxidizing agent. Also remember that the oxidizing agent and reducing agent in a chemical equation must both be reactants.

6. It is essential that you follow a systematic procedure when balancing oxidation-reduction equations. If you learn each step of the method in your textbook and conscientiously do each step for every oxidation-reduction equation you balance, you will probably find that most of your errors are due to a lapse of concentration. Some common errors of this type are (a) forgetting to make sure that other elements are balanced before balancing

hydrogen and oxygen, (b) multiplying only the reactants in a half reaction by the correct factor, or (c) forgetting to simplify the final equation.

The charges on the ions are very important, so be sure to write the charges clearly and neatly. Otherwise it's very easy to overlook an ion in adding up the charges and the balancing will be incorrect.

7. One of the more confusing problem types you will encounter in this chapter is the determination of weight percent by means of a titration. In these problems, the mass of the sample is usually given, and enough information is provided to calculate the mass of some component of that sample. Read the problem carefully to avoid confusing the mass of the sample with the mass of the component.

PRACTICE PROBLEMS

1. (**L2**) Name the following common acids and bases, and indicate whether each is a strong or weak electrolyte.

a. H_2SO_4 _____   b. NH_3 _____

c. H_3PO_4 _____   d. H_2CO_3 _____

2. (**L3**) Rewrite each equation, eliminating spectator ions to create a net ionic equation.

a. $Ca(OH)_2(aq) + HNO_3(aq) \rightarrow Ca(NO_3)_2(aq) + H_2O(\ell)$

b. $Pb(NO_3)_2(aq) + HCl(aq) \rightarrow PbCl_2(aq)^{(s)} + HNO_3(aq)$

3. (**L1**) For each of the following ionic compounds, indicate which are water soluble by indicating the ions formed for the water soluble substances.

a. K_2CO_3 _____   b. $CuCl_2$ _____

c. PbS _____   d. $BaSO_4$ _____

49

4. (**L4**) Complete and balance the following equations. Assume each acid or base is completely neutralized.

a. $KOH + HCl \rightarrow$ _____ + _____

b. $NaOH + H_2SO_4 \rightarrow$ _____ + _____

c. $NH_3 + HBr \rightarrow$ _____ + _____

5. (**L5**) Indicate the acid and base that should be reacted to form each of these salts.

	acid	base
a. $CaCl_2$	_____	_____
b. KNO_3	_____	_____
c. Na_3PO_4	_____	_____
d. $(NH_4)_2SO_4$	_____	_____

6. (**L4**) Complete and balance each of these exchange reactions that is predicted to occur.

a. $Pb(NO_3)_2 + 2 KF \rightarrow$

b. $KOH + (NH_4)_2S \rightarrow$

c. $(NH_4)_2SO_4 + Sr(C_2H_3O_2)_2 \rightarrow$

d. $Fe(NO_3)_3 + 3 KOH \rightarrow$

7. (**L6**) How many liters of 0.444 molar KOH solution can be prepared from 127 grams of KOH?

8. **(L6)** a. What is the molarity of the solution that results when 20.0 mL of water is added to 10.0 mL of 0.200 molar NaOH?
b. What is the molarity of the solution that results when 10.0 mL of a 0.200 molar solution of NaOH is diluted with water to a final volume of 20.0 mL?
c. How much water must be added to 10.0 mL of a 0.200 molar solution of NaOH to produce a final solution that is 0.080 molar in NaOH?

9. **(L7)** How many mL of 0.20 molar barium chloride will be required to react completely with 350 mL of 0.15 molar sulfuric acid according to the equation (balance first if necessary)

$$BaCl_2 \quad + \quad H_2SO_4 \quad \rightarrow \quad BaSO_4 \quad + \quad HCl$$

10. (**L7**) If 27.9 mL of KOH solution is required to titrate 25.0 mL of 0.150 molar HCl to the endpoint, what is the molarity of the KOH solution?

11. (**L7**) Calculate the percent by weight of acetic acid in a vinegar sample if it requires 35.18 mL of 0.420 M NaOH to neutralize the acid in 25.0 mL of a sample of vinegar. You may assume all of the acidity is due to acetic acid, $HC_2H_3O_2$, and that the density of the vinegar is 1.00 g/mL.

12. (**L8**) a 2.00 gram sample of tin(II) chloride is titrated in acidic solution with 0.100 molar potassium permanganate. How many mL of potassium permanganate will be required if the balanced equation for the reaction is

$$16 \ H^+ + 5 \ Sn^{2+} + 2 \ MnO_4^- \ \rightarrow \ 2 \ Mn^{2+} + 5 \ Sn^{4+} + 8 \ H_2O$$

13. (**L8**) Balance the following oxidation-reduction equations.

a. $Mn + Cr^{3+} \rightarrow Mn^{2+} + Cr$

b. $Sn^{2+} + Cr_2O_7^{2-} \rightarrow Cr^{3+} + Sn^{4+}$ (acid solution)

c. $AsO_2^- + ClO^- \rightarrow AsO_3^- + Cl^-$ (basic solution)

PRACTICE PROBLEM SOLUTIONS

1. a. sulfuric acid   strong electrolyte
   b. ammonia   weak electrolyte
   c. phosphoric acid   strong electrolyte
   d. carbonic acid   weak electrolyte

2. a. $OH^- + H_3O^+ \rightarrow 2\ H_2O$

   b. $Pb^{2+} + 2\ Cl^- \rightarrow PbCl_2$

3. a. $K_2CO_3 \rightarrow 2\ K^+ + CO_3^{2-}$

   b. $CuCl_2 \rightarrow Cu^{2+} + 2\ Cl^-$

   c. PbS and d. BaSO$_4$ are insoluble

53

4. a. $KOH(aq) + HCl(aq) \rightarrow KCl(aq) + H_2O(\ell)$

b. $2\ NaOH(aq) + H_2SO_4(aq) \rightarrow Na_2SO_4(aq) + 2\ H_2O(\ell)$

c. $NH_3(aq) + HBr(aq) \rightarrow NH_4Br(aq)$

5. a. $Ca(OH)_2$ and HCl b. KOH and HNO_3

c. $NaOH$ and H_3PO_4 d. NH_3 and H_2SO_4

6. a. $Pb(NO_3)_2(aq) + 2\ KF(aq) \rightarrow PbF_2(s) + 2\ KNO_3(aq)$

b. $KOH + (NH_4)_2S \rightarrow$ no reaction

c. $(NH_4)_2SO_4(aq) + Sr(C_2H_3O_2)_2(aq)$
$$\rightarrow 2\ NH_4C_2H_3O_2(aq) + SrSO_4(s)$$

d. $Fe(NO_3)_3(aq) + 3\ KOH(aq)$
$$\rightarrow Fe(OH)_3(s) + 3\ KNO_3(aq)$$

7. Start with the defining equation for molarity

$$\text{molarity} = \frac{\text{moles of solute}}{\text{volume of solution}}$$

Find the moles of KOH

$$\text{moles KOH} = \frac{\text{grams KOH}}{\text{molar mass}} = \frac{127\ g}{56.11\ g/mol} = 2.263\ mol$$

Rearrange the molarity equation and substitute the known values.

$$\text{liters} = \frac{2.26\ mol}{0.444\ M} = \underline{5.10\ Liters}$$

8. Before you do any of the parts of this problem, read all of them carefully. Notice that even though the wording is very similar, there are significant differences among the three problems. In the first case, you are told how much water is added; the second problem is based on a final volume of solution, and the last part asks how much water should be added. The differences are subtle

but important. Be sure that you can recognize how each type of problem is worked.

a. This is a case where a data table may be helpful.

	initial	final
volume	10.0 mL	10.0 + 20.0 = 30.0 mL
concentration	0.200 M	X

Next substitute into the dilution equation.

$$M_cV_c = M_dV_d$$

$$0.200 \text{ M} \times 10.0 \text{ mL} = X \times 30.0 \text{ mL}$$

$$\underline{X = 0.0667 \text{ M}}$$

b. Again, we will start with a data table.

	initial	final
volume	10.0 mL	20.0 mL
concentration	0.200 M	X

Notice how the changed wording has affected the problem. Next substitute into the dilution equation.

$$M_cV_c = M_dV_d$$

$$0.200 \text{ M} \times 10.0 \text{ mL} = X \times 20.0 \text{ mL}$$

$$\underline{X = 0.100 \text{ M}}$$

c. As before, changing the statement of the problem does affect the data we will use.

	initial	final
volume	10.0 mL	X
concentration	0.200 M	0.080 M

Next substitute into the dilution equation.

$$M_cV_c = M_dV_d$$

$$0.200 \text{ M} \times 10.0 \text{ mL} = 0.080 \text{ M} \times X$$

$$X = 25.0 \text{ mL}$$

But this result is the <u>total</u> final volume, <u>not</u> the amount of water added. To find the amount of water added, it is necessary to subtract the initial volume, 10.0 mL.

$$\text{Water added} = 25.0 \text{ mL} - 10.0 \text{ mL}$$

$$\underline{\text{Water added} = 15.0 \text{ mL}}$$

9. First, write the balanced equation.

$$BaCl_2 + H_2SO_4 \rightarrow BaSO_4 + 2 \text{ HCl}$$

Next, calculate the moles of $BaSO_4$ to be used.

$$0.350 \text{ L H}_2SO_4 \times \frac{0.15 \text{ mol}}{1.00 \text{ L}} = 0.0525 \text{ mol H}_2SO_4$$

Now use the balanced equation to find the mole relationship between barium chloride and sulfuric acid.

$$\text{mol BaCl}_2 = 0.0525 \text{ mol H}_2SO_4 \times \frac{1 \text{ mol BaCl}_2 \text{ required}}{1 \text{ mol H}_2SO_4 \text{ available}}$$

$$= 0.0525 \text{ mol BaCl}_2$$

Finally, calculate the mL of $BaCl_2$ needed.

mol $BaCl_2$ needed =

$$0.0525 \text{ mol BaCl}_2 \text{ required} \times \frac{1.00 \text{ L solution}}{0.20 \text{ mol BaCl}_2}$$

$$= 0.263 \text{ L BaCl}_2$$

or

$$= \underline{260 \text{ mL of BaCl}_2}$$

10. We will follow the same procedure as in the previous problem. First, write the balanced equation.

$$KOH + HCl \rightarrow KCl + H_2O$$

Determine the moles of acid.

$$0.0250 \text{ L HCl} \times \frac{0.150 \text{ mole HCl}}{1.00 \text{ L HCl}} = 0.00375 \text{ mol HCl}$$

Calculate the molarity of base required for 27.9 mL of base to neutralize 0.00375 moles of acid.

$$0.0279 \text{ L NaOH} \times \frac{X \text{ mole NaOH}}{1.00 \text{ L NaOH}} = 0.00375 \text{ mol NaOH}$$

molarity of NaOH = 0.134 M

11. First, write a balanced equation

$$HC_2H_3O_2 + NaOH \rightarrow NaC_2H_3O_2 + H_2O$$

Next, determine how many moles of base have reacted.

$$\text{mol NaOH} = \frac{35.18 \text{ mL}}{1000 \text{ mL/L}} \times \frac{0.420 \text{ mol}}{1.000 \text{ L}}$$

mol NaOH = 0.01478 mol

Next use relationship from the balanced equation to find how many moles of acid would be necessary to react completely with this amount of NaOH.

$$\text{mol } HC_2H_3O_2 = 0.01478 \text{ mol NaOH} \times \frac{1 \text{ mol } HC_2H_3O_2}{1 \text{ mol NaOH}}$$

mol $HC_2H_3O_2$ = 0.01478 moles

Convert the moles to grams

mass of $HC_2H_3O_2$ = 0.01478 mol x 60.06 g/mol

= 0.8874 grams

Since the density of the vinegar is 1.00 g/mL, the mass will be numerically equal to the volume.

mass of vinegar = 25.0 g

The equation for weight percent is

$$\text{weight percent} = \frac{\text{grams of acetic acid} \times 100}{\text{grams of sample (vinegar)}}$$

$$\text{weight percent} = \frac{0.8874 \text{ g} \times 100}{25.0 \text{ grams}}$$

percent of $HC_2H_3O_2$ = 3.55%

12. The balanced equation is already provided.

$$16 \text{ H}^+ + 5 \text{ Sn}^{2+} + 2 \text{ MnO}_4^- \rightarrow 2 \text{ Mn}^{2+} + 5 \text{ Sn}^{4+} + 8 \text{ H}_2\text{O}$$

Start by calculating the moles of $SnCl_2$.

$$\text{moles of } SnCl_2 = \frac{2.00 \text{ g tin(II) chloride}}{189.6 \text{ g/mol}}$$

$$= 0.01055 \text{ mol } SnCl_2$$

Now use the balanced equation to determine how many moles of $KMnO_4$ will be needed.

$$\text{mol } MnO_4^- = 0.01055 \text{ mol } SnCl_2 \times \frac{2 \text{ mol } MnO_4^-}{5 \text{ mol } Sn^{2+}}$$

$$= 0.004219 \text{ mol } MnO_4^-$$

Finally determine the volume of MnO_4^- needed.

$$\text{mol } MnO_4^- = 0.004219 \text{ mol } MnO_4^- \times \frac{1.00 \text{ L solution}}{0.100 \text{ mol } MnO^{4-}}$$

$$= 0.04219 \text{ L } MnO_4^-$$

That is, 42.2 mL of permanganate is required.

13. a. Since the question indicates that these are oxidation-reduction reactions, proceed to Step 2. and write the half reactions.

$$Mn \rightarrow Mn^{2+} \quad \text{(oxidation)}$$

$$Cr^{3+} \rightarrow Cr \quad \text{(reduction)}$$

Step 3. Balance the mass, starting with elements other than hydrogen and oxygen. Both equations are balanced as written.

$$Mn \rightarrow Mn^{2+}$$

$$Cr^{3+} \rightarrow Cr$$

Step 4. Balance the charge by adding electrons to the side of each equation that has the highest positive charge.

$$Mn \rightarrow Mn^{2+} + 2e^-$$

$$Cr^{3+} + 3e^- \rightarrow Cr$$

Notice that the electrons must appear on opposite sides of the two equations. Can you see why this is essential?

Step 5. The manganese reaction produces two electrons each time it occurs, but the chromium reaction requires three electrons each time it occurs. To make the number of electrons produced equal the number required, multiply the manganese equation by three and the chromium reaction by two.

$$3(Mn \rightarrow Mn^{2+} + 2e^-)$$

$$2(Cr^{3+} + 3e^- \rightarrow Cr)$$

Don't forget to multiply the entire equation!

Step 6. Now add the two half reactions together.

$$3\ Mn \rightarrow 3\ Mn^{2+} + 6\ e^-$$

$$2\ Cr^{3+} + 6\ e^- \rightarrow 3\ Cr$$

$$\overline{3\ Mn + 2\ Cr^{3+} \rightarrow 3\ Mn^{2+} + 3\ Cr}$$

Step 7. The mass and charge are balanced on both sides of the equation. In addition, the coefficients cannot be simplified, and none of the species can be canceled as spectator ions. Thus the equation is balanced.

13b. <u>Step 2.</u> The two half reactions are

$$Sn^{2+} \quad \rightarrow \quad Sn^{4+}$$

$$Cr_2O_7^{2-} \quad \rightarrow \quad Cr^{3+}$$

<u>Step 3.</u> Begin by balancing everything <u>except</u> H and O. The other elements are often balanced without changing any coefficients, but always check to make sure. In this case, the chromium is not balanced in the second half reaction.

$$Sn^{2+} \quad \rightarrow \quad Sn^{4+}$$

$$Cr_2O_7^{2-} \quad \rightarrow \quad 2\ Cr^{3+}$$

The tin half reaction is now balanced for mass, so it will not be necessary to do anything further with it until the next step.

Next, balance oxygen. Since the solution is acidic hydrogen ion and water can be added as needed. The reactant side of the chromium half reaction has seven oxygens, the product side none. To balance oxygen, add seven water molecules to the product side.

$$Cr_2O_7^{2-} \quad \rightarrow \quad 2\ Cr^{3+} \quad + \quad 7\ H_2O$$

The last step in balancing the mass is to balance hydrogen by adding hydrogen ions. There are 14 hydrogens on the product side, but none on the reactant side. Therefore add 14 hydrogen ions to the reactant side.

$$14\ H^+ + Cr_2O_7^{2-} \quad \rightarrow \quad 2\ Cr^{3+} \quad + \quad 7\ H_2O$$

<u>Step 4.</u> Balance the charge, adding electrons to the more positive side of each half reaction until both sides have the same charge.

$$Sn^{2+} \quad \rightarrow \quad Sn^{4+} \quad + \quad 2\ e^-$$

$$6\ e^- + 14\ H^+ + Cr_2O_7^{2-} \quad \rightarrow \quad 2\ Cr^{3+} \quad + \quad 7\ H_2O$$

<u>Step 5.</u> Next, multiply by appropriate factors so

60

that the same number of electrons are donated and consumed. The simplest way to do this is to multiply the tin half reaction by 3, so that 6 moles of electrons are donated and six moles are consumed.

$$3 \ (Sn^{2+} \quad \rightarrow \quad Sn^{4+} \quad + \quad 2 \ e^{-})$$

$$6 \ e^{-} + 14 \ H^{+} + Cr_2O_7^{2-} \quad \rightarrow \quad 2 \ Cr^{3+} \quad + \quad 7 \ H_2O$$

Step 6. Now the half reactions are added together to give the overall equation.

$$3 \ (Sn^{2+} \quad \rightarrow \quad Sn^{4+} \quad + \quad 2 \ e^{-})$$

$$6 \ e^{-} + 14 \ H^{+} + Cr_2O_7^{2-} \quad \rightarrow \quad 2 \ Cr^{3+} \quad + \quad 7 \ H_2O$$

$$\overline{14 \ H^{+} + Cr_2O_7^{2-} + 3 \ Sn^{2+} \rightarrow 3 \ Sn^{4+} + 2 \ Cr^{3+} + 7 \ H_2O}$$

Step 7. Notice that the total charge on the reactant side is 18+, and the total charge on the product side is also 18+. Also each side has 14 H, 7 O, 2 Cr, and 3 Sn, so the mass is also balanced.

There are no species on both sides of the equation that can be canceled out, and you cannot divide the coefficients by a constant factor. So the simplest, balanced, ionic equation is

$$14 \ H^{+} + Cr_2O_7^{2-} + 3 \ Sn^{2+} \rightarrow 3 \ Sn^{4+} + 2 \ Cr^{3+} + 7 \ H_2O$$

13c. Step 2. Write the half reactions

$$AsO_2^{-} \quad \rightarrow \quad AsO_3^{-}$$

$$ClO^{-} \quad \rightarrow \quad Cl^{-}$$

Step 3. Balance each half reaction for mass
Arsenic and chlorine are balanced, so move directly to balance oxygen. Count the number of oxygens on each side of the reaction, then add water to the side that has the most oxygens, using one water for each excess oxygen.

$$AsO_2^{-} \quad \rightarrow \quad AsO_3^{-} + H_2O$$

$$H_2O + ClO^{-} \quad \rightarrow \quad Cl^{-}$$

61

Next, on the opposite side from that where you added the water, add two hydroxides for each water molecule that you added.

$$2 \; OH^- + AsO_2^- \rightarrow AsO_3^- + H_2O$$

$$H_2O + ClO^- \rightarrow Cl^- + 2 \; OH^-$$

Step 4. Now balance the half reactions for charge.

$$2 \; OH^- + AsO_2^- \rightarrow AsO_3^- + H_2O + 2 \; e^-$$

$$2 \; e^- + H_2O + ClO^- \rightarrow Cl^- + 2 \; OH^-$$

Step 5. The number of electrons lost in the first equation is the same as the number gained in the second, so it isn't necessary to multiply by a factor.

$$2 \; OH^- + AsO_2^- \rightarrow AsO_3^- + H_2O + 2 \; e^-$$

$$2 \; e^- + H_2O + ClO^- \rightarrow Cl^- + 2 \; OH^-$$

Step 6. Add up the half reactions to give the overall reaction.

$$2 \; OH^- + 2 \; AsO_2^- \rightarrow 2 \; AsO_3^- + H_2O + 2 \; e^-$$

$$2 \; e^- + H_2O + ClO^- \rightarrow Cl^- + 2 \; OH^-$$

$$2 \; OH^- + 2 \; AsO_2^- + H_2O + ClO^- \rightarrow$$
$$Cl^- + 2 \; OH^- + 2 \; AsO_3^- + H_2O$$

Notice that hydroxide and water appear on both sides of the equation. We can simplify the equation by canceling species that appear on both sides of the equation. This means that all of the hydroxides and both of the waters will cancel out, leaving the simplified equation as follows.

$$AsO_2^- + ClO^- \rightarrow Cl^- + AsO_3^-$$

PRACTICE TEST (50 minutes)

1. For each of the following ionic compounds, indicate which ones are water soluble and for the soluble compounds, indicate the ions formed.

a. CaF_2 _____   b. $AgNO_3$ _____

c. $MgCrO_4$ _____   d. $CuSO_4$ _____

2. For each ion or compound determine the oxidation number (with sign) for the underlined element.

a. $H_3\underline{As}O_3$ _____   b. $\underline{P}O^{3-}$ _____

3. Name the following common acids and bases, and indicate whether each is a strong or weak electrolyte.

a. $HC_2H_3O_2$ _____

b. $Ca(OH)_2$ _____

4. Rewrite this equation, eliminating spectator ions, and balance the resulting net ionic equation.

$$K_2S \quad + \quad CdCl_2 \quad \rightarrow \quad CdS \quad + \quad KCl$$

5. Complete and balance equations for the following reactions if they actually occur and tell whether each is an acid-base or a precipitation reaction.

  a. $Na_2SO_4 + Pb(ClO_3)_2 \rightarrow$

  b. $KClO_4 \quad + \quad BeF_2 \quad \rightarrow$

  c. $HC_2H_3O_2 \quad + \quad LiOH \quad \rightarrow$

  d. $CuCl_2 \quad + \quad Na_2S \quad \rightarrow$

6. Suppose that you wish to prepare 265 mL of 0.500 molar sulfuric acid. How many grams of H_2SO_4 would be required to prepare this solution?

7. How many mL of 0.200 molar HCL would be necessary to exactly neutralize 5.00 grams of $Ba(OH)_2$?

63

8. a. How much water must be added to 2.50×10^2 mL of 2.60 M HCl to produce a 0.500 M solution of HCl?
b. Calculate the molar concentration of an HCl solution formed by adding 120.0 mL of water to 10.0 mL of 5.20 M HCl.

9. If sodium metal is placed in water it will react according to the balanced equation

$$2\ Na(s) + 2\ H_2O(\ell) \rightarrow 2\ NaOH(aq) + H_2(g)$$

Calculate the molar concentration of sodium hydroxide formed in a reaction of this type if it is known that 0.020 grams of hydrogen was generated and the total volume of water after the reaction was 150 mL.

10. A 1.46 gram sample of iron ore is dissolved in acid and reduced so that all of the iron present is in the +2 oxidation state. The resulting solution is titrated to the endpoint with 19.8 mL of 0.31 molar cerium(IV) ion according to the equation (not balanced)

$$Fe^{2+}\ +\ Ce^{4+}\ \rightarrow\ Fe^{3+}\ +\ Ce^{2+}$$

What is the percent of iron in the original sample?

11. Write balanced equations for the preparation of each compound.

a. NH_4NO_3 b. $CaCl_2$ c. $CuCO_3$ d. $Ba(NO_3)_2$

12. Complete and balance this oxidation-reduction equation in acidic solution.

$$C_2O_4^{2-}\ +\ Cr_2O_7^{2-}\ \rightarrow\ Cr^{3+}\ +\ CO_2\ +\ H_2O$$

CONCEPT TEST ANSWERS

1. nonelectrolyte   2. strong
3. spectator ions   4. base
5. $OH^-\ +\ H_3O^+\ \rightarrow\ 2\ H_2O$
6. moles, moles
7. equivalence point (neutralization point or end point is also acceptable.)

8. indicator
9. standardization
10. electrons
11. reduced, reducing
12. oxidation number
13. a. Ti +4, Cl -1   b. S +6, O -2
    c. H +1, Cl +7, O -2   d. Si +4, H -1
14. and 15.
  For equation a
      element oxidized: oxygen (-2 to 0)
      element reduced: chlorine (0 to -1)
      oxidizing agent: Cl_2 reducing agent: H_2O
For equation b
      element oxidized: tin (0 to +2)
      element reduced: nitrogen (+5 to +2)
      oxidizing agent: HNO_3 reducing agent: Sn

PRACTICE TEST ANSWERS

1. (L1) b. $AgNO_3(aq) \rightarrow Ag^+(aq) + NO_3^-(aq)$

   d. $CuSO_4(aq) \rightarrow Cu^{2+}(aq) + SO_4^{2-}(aq)$

   a. CaF_2 and c. $MgCrO_4$ are insoluble

2. (L8) a. +3 b. +5
3. (L2) a. acetic acid   weak electrolyte
       b. calcium hydroxide   strong electrolyte

4. (L3) $S^{2-}(aq)$ + $Cd^{2+}(aq) \rightarrow$ CdS(s)

5. (L3)a. $Na_2SO_4(aq) + Pb(ClO_3)_2(aq)$

                     $\rightarrow PbSO_4(s) + 2 NaClO_3(aq)$

   b. $KClO_4$ + BeF_2 $\rightarrow$ no reaction

   c. $HC_2H_3O_2(aq) + LiOH(aq) \rightarrow LiC_2H_3O_2(aq) + H_2O(\ell)$

   d. $CuCl_2(aq)$ + $Na_2S(aq) \rightarrow CuS(s)$ +   2 NaCl(aq)

   Reaction d is acid-base, the other two are
   precipitation reactions.
6. (L6) 13.0 grams
7. (L7) 292 mL
8. (L6) a. 1050 mL b. 0.400 M
9. (L6) 0.13 M
10. (L8) 47%

11. (L5) Since a number of different answers are acceptable in each case, your answer will probably not be the same as that given. For your answer to be correct, however, both of the reactants must be soluble, and at least one of the products must be insoluble.

a. $2 AgNO_3(aq) + (NH_4)_2S(aq) \rightarrow 2 NH_4NO_3(aq) + Ag_2S(s)$

b. $CaSO_4(aq) + BaCl_2(aq) \rightarrow CaCl_2(aq) + BaSO_4(s)$

c. $Cu(NO_3)_2(aq) + Na_2CO_3(aq) \rightarrow CuCO_3(s) + 2 NaNO_3(aq)$

d. $BaCl_2(aq) + 2 AgNO_3(aq) \rightarrow Ba(NO_3)_2(aq) + 2 AgCl(s)$

12. (L8) $3 C_2O_4^{2-} + Cr_2O_7^{2-} + 14 H^+$

$$\rightarrow 2 Cr^{3+} + 6 CO_2 + 7 H_2O$$

SPECIAL SECTION I
NORMALITY AND EQUIVALENTS

In some cases when the same type of stoichiometry calculation is done repeatedly, it is especially convenient to use a concentration unit called normality. The idea behind this concentration unit is that if the amount of each reactant is measured relative to a set amount of some constant, standard reactant, it will be easier to calculate the stoichiometric relationships. This standard amount is called a *chemical equivalent*, because it provides a way to identify quantities of reactants that will have equivalent chemical reactivities.

As will be seen, this approach does tend to make the easier problems a little harder, but it also can help to simplify some rather difficult problems. Since normality is not defined in exactly the same way for acid-base reactions as for oxidation-reduction reactions, these two situations will be discussed separately.

SECTION 1. CHEMICAL EQUIVALENTS AND ACID-BASE TITRATIONS

According to one of the most commonly used acid-base definitions, an acid is a substance that donates protons, that is hydrogen ions, and a base is a substance that accepts protons. This suggests that the amount of acids and bases can be related to the ability to produce or react with one mole of hydrogen (or hydronium) ions. This is the basis of the definition of the chemical equivalent in acid-base reactions.

For acid-base reactions, a **chemical equivalent** *is defined as the amount of an acid or base that will react with or produce one mole of hydrogen ions.*

For most common acids, the number of equivalents provided by each mole of compound is determined from the ionization equation. For example, compare hydrochloric acid and sulfuric acid:

$$HCl \longrightarrow H^+ + Cl^-$$

$$H_2SO_4 \longrightarrow 2 H^+ + SO_4^{2-}$$

One mole of hydrochloric acid will produce one mole of hydrogen ions, or *one equivalent*. Thus, for HCl there is one equivalent per mole. One mole of sulfuric acid will produce two moles of hydrogen ions and so has two equivalents per mole.

It is also easy to determine the number of equivalents per mole for most common bases. Since the neutralization reaction between hydrogen ion and hydroxide ion requires one mole of hydrogen ion for each mole of hydroxide,

$$H^+(aq) + OH^-(aq) \longrightarrow H_2O(\ell)$$

one chemical equivalent of a base is the amount that will produce one mole of hydroxide ions.

As before, in order to determine equivalents per mole for bases it is necessary to examine the ionization equations. For example, compare sodium hydroxide and calcium hydroxide.

$$NaOH \longrightarrow Na^+ + OH^-$$

$$Ca(OH)_2 \longrightarrow Ca^{2+} + 2 OH^-$$

The number of equivalents per mole for bases is determined by the number of moles of hydroxide ion produced per mole of base. For sodium hydroxide, there is one equivalent per mole, and for calcium hydroxide, there are two equivalents per mole.

It has probably become apparent that one way to find equivalents per mole is to simply count the number of ionizable hydrogens or hydroxides in the formula of the compound. Frequently this works, but there are a number of cases, like ammonia, NH_3, where it's essential to refer to the ionization equation when determining equivalents per mole.

$$NH_3 \quad + \quad H_2O \quad \rightarrow \quad NH_4^+ \quad + \quad OH^-$$

From the equation it should be easy to see that ammonia has one equivalent per mole.

Once the number of equivalents per mole for a compound has been determined, usually the next step is to calculate the equivalent mass. *The* **equivalent mass** *of a compound is the mass of that compound needed to produce one chemical equivalent.* Some instructors may call this quantity the gram equivalent mass, equivalent weight or may simply use the units, grams/equivalent. The number of equivalents per mole provides an easy conversion to equivalent mass from molar mass.

$$\text{equivalent mass} = \frac{\text{molar mass}}{\text{equivalents/mole}}$$

It is important to note at this time that the equivalents per mole for a compound (and therefore also the equivalent mass) is determined by its behavior in a given chemical reaction. This doesn't present many problems in acid-base chemistry, but it does create some difficulty for oxidation-reduction reactions. Examples of this situation will be discussed later in this chapter.

The preceding calculation determined that 36.46 grams of HCl represents one equivalent in typical acid-base reactions. If the mass of HCl is greater than or less than 36.46 grams, obtain the number of equivalents by dividing the available mass by 36.46 g/eq. Notice that equivalents are not the same as equivalents per mole. The only limits on the number of equivalents of HCl are set by the ability to obtain different masses of this compound, but the number of equivalents per mole for HCl is established by the ionization equation. In order to change the eq/mol for HCl it would be necessary to change the ionization equation.

69

EXAMPLE 1. CALCULATION OF EQUIVALENTS

Calculate the number of equivalents in a 2.00 gram sample of (a) HCl and (b) $Ba(OH)_2$

(a) As seen above, the ionization of HCl produces one mole of hydrogen ions per mole of HCl or one eq/mol. Therefore, the equivalent mass is numerically equal to the molar mass.

$$\text{equivalent mass} = \frac{36.46 \text{ g/mol}}{1 \text{ eq/mol}} = 36.46 \text{ g/eq}$$

Divide the mass of sample by the equivalent mass to obtain equivalents

$$\text{eq of HCl} = \frac{2.00 \text{ g}}{36.46 \text{ g/eq}} = 0.0549 \text{ eq}$$

(b) The ionization equation

$$Ba(OH)_2 \rightarrow Ba^{2+} + 2 \ OH^-$$

indicates two equivalents per mole for this compound. Using the standard equation to calculate the equivalent mass

$$\text{equivalent mass} = \frac{171.3 \text{/mol}}{2 \text{ eq/mol}} = 85.66 \text{ g/eq}$$

As in the previous example, the sample mass is divided by the equivalent mass.

$$\text{eq of } Ba(OH)_2 = \frac{2.00 \text{ g}}{85.66 \text{ g/eq}} = 0.0233 \text{ eq}$$

Since most of the acid-base reactions that will be of concern occur in solution, it is necessary to create a concentration unit based on equivalents. The new unit, called normality, is analogous to the molar concentration unit explained

70

previously. Just as molarity is defined as moles of solute per liter of solution, **normality** *is defined as equivalents of solute per liter of solution*. The units of normality are eq/liter, but it is common to use N as a symbol for this unit. Another way to think about normality is that it is simply the number of moles per liter of hydrogen ion or hydroxide ion in a solution.

Notice that since the number of equivalents per mole is always greater than a value of one for acids and bases, one mole must always represent one or more equivalents. This means that normality can be equal to or greater than molarity but never less than the molarity.

Another way to represent the relationship between normality and molarity is

$$normality = molarity \times \frac{equivalents}{mole}$$

This equation is very useful for converting from one of these concentration units to another. To prove that it is correct, use dimensional analysis.

$$\frac{equivalents}{liter} = \frac{moles}{liter} \times \frac{equivalents}{mole}$$

Notice that when moles are canceled, the result is, indeed, an equality.

EXAMPLE 2. CONVERTING MOLARITY TO NORMALITY

Calculate the normality of a 3.0 Molar solution of H_3PO_4.

<u>Solution</u>

First, write the ionization reaction for H_3PO_4 and determine the number of equivalents per mole.

$$H_3PO_4 \rightarrow 3\ H^+ + PO_4^{3-}$$

There are three equivalents per mole for H_3PO_4.

Next use the relationship between normality and molarity developed above.

$$\text{normality} = \text{molarity} \times \frac{\text{equivalents}}{\text{mole}}$$

Substituting,

normality = 3.0 Molar x 3 eq/mol   = 9.0 N

normality = 9.0 N

Thus a 3.0 Molar solution of H_3PO_4 has a normality of 9.0 Normal in typical acid-base reactions.

SECTION 2. USING EQUIVALENTS AND NORMALITY IN ACID-BASE TITRATIONS

The preceding section explained that equivalents are a measure of the amount of acid or base in terms of the ability to produce or react with hydrogen ions. One equivalent of any acid will produce one mole of hydrogen ions, and one equivalent of any base will react with one mole of hydrogen ions. It follows that one equivalent of any acid is required to react completely with one equivalent of any base. This is the definition of the equivalence point for any acid-base titration. It can also be stated in the form of an equation. At the **equivalence point** of an acid-base titration,

Equivalents of acid = Equivalents of base

Chemists also often talk about the end point of a titration, and it is reasonable to ask how it is related to the equivalence point. The **end point** of a titration occurs when sufficient titrant has been added to cause a chemical indicator to change color. Since these chemical indicators don't change at exactly the point where the equivalents are equal, the end point is not exactly the same as the equivalence point. The two values are usually close enough, however, so normally the value of the end point and the equivalence point are presumed to be the same.

The statement that equivalents of acid equals equivalents of base is the starting point for most of the problems in this section. Almost all of the equations required for acid-base titrations can be obtained by substituting a few simple relationships into this relationship based on equivalents. For example, the equation

normality = equivalents/liter.

can easily be rearranged to the form

normality x Volume (in liters) = equivalents.

It has also been shown earlier that

equivalents = grams/equivalent mass.

Depending on what information is available, either of these can be substituted for equivalents of acid or base in the original equation. Most titration problems can be worked with some combination of these three equations.

Students often complain that there are too many equations to memorize when doing acid-base calculations with equivalents. There appear to be many different equations, but they are not all independent. Actually most of them are obtained by simple substitutions, such as those described in the previous paragraph. Once this is recognized, it's easy to design equations for each problem that is encountered. Thus, this is an important technique to master.

EXAMPLE 3. USING EQUIVALENTS IN AN ACID-BASE TITRATION

If 25.84 mL of 0.120 N HCl is required to titrate 100.0 mL of barium hydroxide to the equivalence point, what is the normality of the $Ba(OH)_2$?

Solution

The first piece of important information is that the calculation deals with a titration at the equivalence point. Therefore, begin with the equation

$$\text{eq of } Ba(OH)_2 = \text{eq of } HCl$$

Since the normality and the volume of the hydrochloric acid is given, it is reasonable to substitute eq = N x V on the hydrochloric acid side of the equation. The volume of the barium hydroxide is given and the normality is the unknown, so also insert eq = N x V on that side of the equation.
The resulting equation is

$$N\ (Ba(OH)_2)\ x\ V\ (Ba(OH)_2) = N\ (HCl)\ x\ V\ (HCl)$$

Notice that strictly speaking both of these volumes should be in liters. In this case, since the same conversion factor will occur on both sides of the equality sign, it will cancel out. This means that the correct answer would result even if the volumes are not converted to liters. This is not always true. The next example is a case where failure to convert to liters will produce an incorrect answer. It's never wrong to convert all volumes to liters in these problems, so for students who are unsure, the best strategy is always to make the conversion.

Now, inserting the known values

$$N\ (Ba(OH)_2)\ x\ 0.1000\ L = 0.120\ N\ x\ 0.02584\ L$$

$$N\ (Ba(OH)_2) = \underline{0.0310\ N}$$

EXAMPLE 4. USING EQUIVALENTS WITH A SOLID REACTANT

How many grams of solid NaOH are required to neutralize 250.0 mL of 2.00 N H_3PO_4?

Solution

The language in this problem is somewhat different, since it doesn't specifically refer to the equivalence point. It does say that an acid and a base are being reacted until they neutralize each other, and this occurs at the equivalence point. Thus, we can use the same relationship.

eq of base (NaOH) = eq of acid (H_3PO_4)

Since the volume and normality of the H_3PO_4 are given, substitute eq = N x V on the right hand side of the equation. Neither normality nor volume is available for the NaOH, and so a different relationship is needed. To find grams, use the relationship that eq = grams /equivalent mass. Is enough information provided to determine the equivalent mass for NaOH? Yes!

This completes the left hand side of the equation.

$$\frac{\text{grams (NaOH)}}{\text{equivalent mass (NaOH)}} = N\ (H_3PO_4)\ x\ V\ (H_3PO_4)$$

Now calculate the equivalent mass for NaOH. This compound has one equivalent per mole, so the equivalent mass is numerically the same as the molar mass, or 40.0 g/mol.

Notice that in this problem failure to convert the volume to liters will produce a large error in the answer. For this reason it is essential to convert the volume before substituting the values in the above equation.

$$\frac{\text{grams (NaOH)}}{40.00\ \text{g/eq}} = 2.00\ N\ x\ 0.250\ L$$

grams (NaOH) = 2.00 N x 0.250 L x 40.0 g/eq

grams (NaOH) = 20.0 grams

There is one problem type that may cause confusion when using the above method to solve titration problems. If the mass of the sample is given and the percentage of some component in that sample is the unknown, it's tempting to use the sample mass directly in the equivalents relationship. This is incorrect. The sample consists of at least two components, an acid or base that is being titrated, and some inert material that does not participate in the reaction. If this is the case, the mass provided cannot be substituted into the equivalents relationship, since it does not represent the amount of acid or base.

EXAMPLE 5. CHEMICAL ANALYSIS USING EQUIVALENTS

Fumaric acid, $C_4H_4O_4$, is a substance essential for vegetable and animal respiration. It can produce two moles of ionizable hydrogen ions per mole of acid. Determine the weight percent of fumaric acid in a sample that is a mixture consisting only of fumaric acid and a solid, inert material if it requires 25.35 mL of 0.650 N NaOH to titrate a 2.010 gram sample of this mixture to the equivalence point?

Solution

This titration has gone to the equivalence point, so the first step is to write the equation

eq of fumaric acid = eq of sodium hydroxide

The normality and the volume of the sodium hydroxide are given, and so it is obvious that N (NaOH) x V (NaOH) can be substituted for eq of NaOH on the right hand side of the equation. As was just noted, the 2.010 gram mass is not fumaric acid, but a mixture; however, the mass of fumaric acid in that mixture is needed.

Since the molar mass and the equivalents per mole are given for fumaric acid, calculate the equivalent mass of this substance. Then, on the left side of the equation, substitute grams divided by equivalent mass for equivalents of fumaric acid.

76

First, calculate the equivalent mass of fumaric acid

$$\text{equivalent mass} = \frac{\underline{\text{molar mass}}}{\text{eq/mol}}$$

$$\text{equivalent mass} = \frac{116.08 \text{ g/mol}}{2 \text{ eq/mol}} = 58.04 \text{ g/eq}$$

Next, make the appropriate substitutions in the equivalents relationship

$$\frac{\underline{\text{gram of fumaric acid}}}{\text{eq mass of fumaric acid}} = \text{N (NaOH)} \times \text{V (NaOH)}$$

Convert the volume to liters, insert the known values and solve for grams of fumaric acid.

$$\frac{\underline{\text{grams of fumaric acid}}}{58.04 \text{ g/eq}} = 0.650 \text{ N} \times 0.02535 \text{ L}$$

$$\text{grams of fumaric acid} = 0.9564 \text{ grams}$$

Finally, determine the weight percent of fumaric acid in the sample.

$$\text{weight percent} = \frac{\text{grams of fumaric acid} \times 100}{\text{grams of sample}}$$

$$\text{weight percent} = \frac{0.9564 \text{ g} \times 100}{2.010 \text{ g}}$$

$$\underline{\text{weight percent} = 47.6\%}$$

Notice that in each case, the essential relationship has been the observation that the equivalents of acid must equal the equivalents of base. This must be true at the equivalence point of any titration according to the definition. In some problems there may be more than one acid or base. That will not require any change in the approach. At the equivalence point, the total number of equivalents of all of the acids present still must be identical with the total number of equivalents of all of the bases present. If there are two different acids used to titrate a base to the equivalence point, the equivalents relationship becomes

eq of acid 1 + eq of acid 2 = eq of base

and the same equations are substituted for equivalents as in the previous examples.

SECTION 3. USING EQUIVALENTS FOR OXIDATION-REDUCTION REACTIONS

Chemical equivalents can also be used for oxidation-reduction reactions, but some change is needed, since hydrogen ions are no longer the best measure of chemical reactivity. In the previous study of oxidation-reduction processes, it was pointed out that the transfer of electrons was the essential component in these reactions. This suggests that a different definition of the chemical equivalent is appropriate.

For oxidation-reduction reactions, a **chemical equivalent** *is defined as the amount of substance that will react with or produce one mole of electrons.*

It is rarely possible to determine equivalents per mole for oxidizing agents or reducing agents by inspection of either the formula of the substance or the balanced equation. *Equivalents per mole for oxidizing agents or reducing agents are determined from the balanced half-reactions.* Unlike acid-base problems, there are few shortcuts here; however, once it is recognized that the equivalents per mole cannot be determined by inspection of the formula, the process is not really too difficult.

It has already been pointed out in Chapter 5 of the textbook that *loss of electrons is oxidation and gain of electrons is reduction.* Also, the substance oxidized in a chemical reaction is the *reducing agent*, and the substance reduced is the *oxidizing agent.* This is also a good time to review the method for balancing equations by half-reactions, since these techniques will be used again in this section.

In the left-hand column below are a list of balanced half-reactions. The right-hand column lists the number of eq/mol for the reactant in each half-reaction. Cover the right-hand column and try to determine the eq/mol without looking.

half-reaction	eq/mol
$Cu^{2+} + 2\,e^- \rightarrow Cu$	2
$Fe^{2+} \rightarrow Fe^{3+} + e^-$	1
$MnO_4^- + 8\,H^+ + 5\,e^- \rightarrow Mn^{2+} + 4\,H_2O$	5
$2\,Cl^- \rightarrow Cl_2 + 2\,e^-$	1

Notice that in the last reaction, two moles of electrons are produced, but this also requires two moles of chloride ions. Thus there are <u>two</u> equivalents for <u>two</u> moles of chloride, or one equivalent per mole.

Aside from the method for determining equivalents per mole, most of the other techniques necessary to apply equivalents to oxidation-reduction reactions are very similar to those previously learned. Terms such as normality, equivalent mass, and equivalence point still have the same meaning. The only important difference is the change in the method of determining equivalents per mole.

EXAMPLE 6. USING EQUIVALENTS FOR OXIDATION-REDUCTION REACTIONS

Calculate the normality of a certain potassium permanganate, $KMnO_4$, solution if 26.90 mL of this solution is required to titrate 1.00 gram of iron(II) to the equivalence point in acidic solution. The skeleton equation (not balanced) for this process is

$$Fe^{2+} \quad + \quad MnO_4^- \quad \rightarrow \quad Fe^{3+} \quad + \quad Mn^{2+}$$

Now calculate the molarity from the normality.

<u>Solution</u>

As was the case in acid-base problems, the first step is to recognize that

eq of oxidizing agent = eq of reducing agent

Iron is the reducing agent, since it is oxidized from +2 to +3. Permanganate ion is the oxidizing agent, since manganese is reduced from +5 to +2.

Next determine what to substitute into the equivalents relationship. The mass of iron is given, and the equivalent mass of iron can be determined from the information given. It is reasonable, then, to substitute grams divided by equivalent mass for equivalents of reducing agent. The volume of permanganate is given, and the normality is the unknown, so substitute N x V for equivalents of oxidizing agent. (Don't forget that the volume must be in liters.)

Before making these substitutions, determine the equivalent mass for iron(II). From the skeleton equation, iron(II) is oxidized to iron(III), and the balanced half-reaction is easy to write:

$$Fe^{2+} \quad \rightarrow \quad Fe^{3+} \quad + \quad e^-$$

This indicates that there is one equivalent per mole for iron(II). Since iron(II) is an ion, not a molecule, the equivalent mass is obtained by dividing the <u>atomic weight</u> by the equivalents per mole

$$\text{equivalent mass} \quad = \quad \frac{\text{atomic weight}}{\text{eq/mole}}$$

80

$$\text{equivalent mass} = \frac{55.85 \text{ g/mol}}{1 \text{ eq/mole}}$$

$$\underline{\text{equivalent mass} = 55.85 \text{ g/eq}}$$

Now substitute into the equivalents relationship.

$$N \text{ (MnO}_4^-) \times V \text{ (MnO}_4^-) = \frac{\text{grams Fe}^{2+}}{\text{eq mass of Fe}^{2+}}$$

$$N \text{ (MnO}_4^-) \times .0269 \text{ L} = \frac{1.00 \text{ g}}{55.85 \text{ g/eq}}$$

$$\underline{N \text{ (MnO}_4^-) = 0.666 \text{ N}}$$

There are several different ways to determine the molarity of this solution but probably the simplest is to use the equation

normality = molarity x eq/mol

To determine the eq/mol for permanganate, it's necessary to use the balanced half-reaction. From the skeleton reaction given, the basic components are

$$\text{MnO}_4^- \rightarrow \text{Mn}^{2+}$$

Balancing this according to the procedure from Chapter 5 produces

$$\text{MnO}_4^- + 8 \text{ H}^+ + 5 \text{ e}^- \rightarrow \text{Mn}^{2+} + 4 \text{ H}_2\text{O}$$

This indicates that there are five eq/mol for MnO_4^-. Substituting this value and the normality into the original equation

$$0.666 \text{ N} = \text{molarity} \times 5 \text{ eq/mol}$$

$$\underline{\text{molarity} = 0.133 \text{ M}}$$

In the previous problem, the reactants were all ionic. In many cases, the amount of the reactants is given in terms of the compound which serves as a source of the ion. For instance, what if the preceding problem had been stated in terms of 1.00 gram of iron(II) chloride rather than 1.00 gram of iron(II)? Under these circumstances, the same method would still be used to determine eq/mol, but the equivalent mass would change. Instead of obtaining the equivalent mass by dividing the atomic weight of iron by 1 eq/mol, divide the molar mass of iron(II) chloride by 1 eq/mol. Otherwise, the set-up would remain the same.

It is important to realize that the equivalent mass of a species depends on the specific reaction in which it is used. For example, suppose that the reaction described in Example 6 were done in basic solution rather than acidic solution. The product of the permanganate reduction would then be manganese(IV) oxide rather than manganese(II) ion. The permanganate half-reaction is now

$$MnO_4^- \ + \ 2\ H_2O \ + \ 3\ e^- \ \rightarrow \ MnO_2 \ + \ 4\ OH^-$$

It is clear that permanganate now has <u>three</u> equivalents per mole. This means that even if the permanganate for the titration is taken from exactly the same stock solution, it will have a different normality, because the equivalents per mole will be different.

In order to define the normality of a solution, it is <u>essential</u> to know the equation for the reaction in which the solution is to be used. If a change in the conditions alters that equation, this will change the normality of the solution. Usually this is best accomplished by converting the old normality to molarity, then determining the new normality based on that molarity. Changing the reaction doesn't change the molarity, so it provides a constant value for comparison with the normality.

EXAMPLE 8. CALCULATING A NEW NORMALITY WHEN EQUIVALENTS/MOLE CHANGES

The normality of a certain solution of potassium permanganate is determined to be 6.0 Normal in acidic solution where the half reaction is

$$MnO_4^- + 8 H^+ + 5 e^- \rightarrow Mn^{2+} + 4 H_2O$$

Calculate the normality of that same solution when the potassium permanganate is to be used in basic solution where the half-reaction is

$$MnO_4^- + 2 H_2O + 3 e^- \rightarrow MnO_2 + 4 OH^-$$

Solution

First, convert the original normality to molarity. From the half-reaction, it is clear that in acidic solution there are five equivalents per mole for permanganate. Substituting the available information into the equation that relates normality and molarity

normality = molarity x eq/mol

6.0 N   = molarity x 5 eq/mol

molarity = 1.2 M

Next, examination of the half-reaction for basic conditions indicates that there are now three eq/mol. Use the same relationship as before to convert the molarity to the new normality.

normality = molarity x eq/mol

normality = 1.2 M x 3 eq/mol

normality = 3.6 N

It has been shown that it's slightly more difficult to set up titration problems using normality that it would be using molarity. Determining equivalents per mole is not much harder than using the balanced equation to find a stoichiometric relationship, but using equivalents does require some new concepts. On the other hand, when doing the same titration over and over again, it is easier to use normality, especially if several reactants must be related to each other. *Regardless of whether normality or molarity is used for these problems, it is important to recognize that both methods represent the same basic chemical concepts.*

Study Questions for Chemical Equivalents

1. Barium hydroxide reacts with HCl according to the balanced equation

$$Ba(OH)_2(aq) + 2\ HCl(aq) \rightarrow BaCl_2(aq) + 2\ H_2O(\ell)$$

(a) If 1.7 grams of barium hydroxide are dissolved in enough water to produce 125 mL of solution, what is the molarity of the resulting solution?   (b) What is the normality of the solution? (c) What is the molar concentration of OH⁻ in this solution?

2. Succinic acid, $C_4H_6O_4$, loses two hydrogen ions on reaction with a base. If 2.50 grams of this acid is dissolved in enough water to produce 255 mL of solution, (a) what is the normality of the resulting solution?   (b) What is the molar concentration of H^+ in this solution?

3. Calculate the normality of a sulfuric acid solution prepared by dissolving 25.0 grams of H_2SO_4 in enough water to produce 150. mL of solution. What is the molarity of this solution? What is the concentration of H^+ in this solution?

4. (a) How many equivalents of acid are contained in 300.0 mL of 0.15 N oxalic acid, $H_2C_2O_4$?   (b) If this solution is diluted to a volume of 500.0 mL,

84

what is the new normality? (c) Did the number of equivalents of acid change when the solution was diluted?

5. (a) How many milliliters of 0.250 N HCl would be required to neutralize completely 2.50 grams of NaOH? (b) How many milliliters of 0.250 N H_2SO_4 would be require to neutralize the same mass of NaOH?

6. How many milliliters of 0.110 N HCl will be required to titrate 5.00 grams of $Ba(OH)_2$ to the equivalence point?

7. Calculate the equivalent mass of a certain unknown acid if 31.0 mL of 0.132 N NaOH is require to titrate a 0.500 gram sample of the unknown acid to the equivalence point.

8. A 1.00 gram sample of citric acid, $C_6H_8O_7$, requires 31.23 mL of 0.500 N NaOH for titration to the equivalence point. (a) What is the equivalent mass of citric acid? (b) What is the number of equivalents per mole for citric acid?

9. What is the percent by weight of acetic acid in a vinegar sample if it requires 30.24 mL of 0.210 N KOH to neutralize the acid in 10.0 mL of vinegar. You may assume all of the acidity is due to acetic acid, $HC_2H_3O_2$, and that the density of the vinegar is 1.00 g/mL.

10. Suppose that you are attempting to determine the equivalent mass of a substance commonly used in antacid tablets. You dissolve 0.210 grams of the pure material (which is a base) in 25.0 mL of 0.400 N HCl and then titrate the resulting mixture with 0.110 N NaOH. If 18.2 mL of NaOH are required to reach the end point, what is the equivalent mass of the antacid?

11. A solution is prepared by mixing 31.21 mL of 0.100 N HCl with 98.53 mL of 0.500 N H_2SO_4 and then adding 50.0 mL of 1.002 N $Ca(OH)_2$ to the resulting mixture. (a) Is the resulting solution acidic or basic? (b) How many milliliters of 0.300 N acid or

base (as appropriate) must be added to the solution to make it exactly neutral?

12. The skeleton equation (not balanced) for the reduction of iron(II) with tin(II) is as follows:

$$Fe^{3+} \quad + \quad Sn^{2+} \quad \rightarrow \quad Fe^{2+} \quad + \quad Sn^{4+}$$

Answer the following questions based on this equation. (a) What is the equivalent mass of tin(II) chloride? (b) How many grams of $SnCl_2$ must be dissolved in 2.00 liters of water to produce a 0.200 N solution of tin(II)? (c) How many equivalents of iron(III) are present in a 5.00 gram sample of iron(III) chloride? (d) How many milliliters of 0.200 N tin(II) solution will be required to exactly reduce 5.00 grams of iron(III) chloride to iron(II) chloride?

13. The equation for the reaction of hydrogen sulfide with iodine is

$$H_2S \quad + \quad I_2 \quad \rightarrow \quad 2 \, I^- \quad + \quad 2 \, H^+ \quad + \quad S$$

According to this reaction, (a) How many grams of I_2 are contained in 0.750 liters of a 0.330 N iodine solution? (b) How many milliliters of 0.330 N iodine solution will be required to react completely with 0.225 grams of H_2S?

14. The skeleton equation (not balanced) for the reaction of iron(II) with permanganate ion in acidic solution is as follows:

$$MnO_4^- \quad + \quad Fe^{2+} \quad \rightarrow \quad Mn^{2+} \quad + \quad Fe^{3+}$$

Answer the following questions based on this equation.
(a) Calculate the number of equivalents in 25.0 mL of 0.110 N MnO_4^-. (b) How many grams of $KMnO_4$ must be dissolved in 250. mL of water to produce a 0.110 N solution? (c) Calculate the number of equivalents in 2.00 grams of $FeCl_2$. (d) How many milliliters of 0.110 N $KMnO_4$ will be required to titrate a 2.00 sample of pure iron(II) chloride to the equivalence point?

86

15. When permanganate ion acts as an oxidizing agent in basic solution, the product is MnO_2, as shown in the balanced half-reaction,

$$MnO_4^- \ + \ 2 \ H_2O \ + \ 3 \ e^- \ \rightarrow \ MnO_2 \ + \ 4 \ OH^-$$

The solution of potassium permanganate described in the preceding question had a normality of 0.110 when it was used in acidic reactions. What would be the normality of this same permanganate when reacting in basic solution to form MnO_2, as shown in the above half-reaction?

16. The compound $Fe(NH_4)_2(SO_4)_2 \cdot 6H_2O$ (molar mass = 392.19 g/mol) is sometimes used to standardize permanganate solutions. If 32.45 mL of aqueous $KMnO_4$ is required to titrate the iron(II) ion in a 1.050 gram sample of this compound to the equivalence point, what is the normality of the $KMnO_4$? The balanced equation for the reaction is

$$MnO_4^- \ + \ 5 \ Fe^{2+} \ + \ 8 \ H^+ \ \rightarrow \ Mn^{2+} \ + \ 5 \ Fe^{3+} \ + \ 4 \ H_2O$$

17. A sample of iron ore weighing 2.010 grams was dissolved in acid, and the iron was all reduced to soluble iron(II). What is the weight percent of iron in the ore sample if 34.7 mL of 0.142 N $KMnO_4$ is required to titrate the iron(II) to the equivalence point? The balanced equation for this reaction is given in the previous problem.

18. Zinc oxide is used as a pigment in white paints. It is basic and will react with acids according to the equation

$$ZnO \ + \ 2 \ H^+ \ \rightarrow \ H_2O \ + \ Zn^{2+}$$

A 2.820 gram sample of technical grade (i.e. not pure) zinc oxide is dissolved in 50.0 mL of 1.00 N sulfuric acid and not all of the acid is neutralized. The resulting solution is then titrated with 0.137 N KOH, and 3.30 mL of KOH is required to reach the equivalence point. What is the weight percent of ZnO in the original sample? (Assume that the sample contains no basic material other than the ZnO.)

19. A solution of nitric acid, HNO_3, is prepared to be 4.00 N when acting as an acid. Later on it is decided to use this solution in a redox reaction that involves the reduction of nitric acid to NO. How many mL of water must be added to a 75.0 mL sample of the original nitric acid solution in order to make it 3.00 N when acting as an oxidizing agent?

STUDY QUESTION ANSWERS

1. a. 0.079 M b. 0.160 N c. 0.160 M
2. a. 0.166 N b. 0.166 M
3. 3.40 N, 1.70 M, 3.40 M
4. a. 0.045 equivalents b. 0.090 N c. no
5. a. 250 mL b. 250 mL
6. 531 milliliters
7. 122 g/eq
8. a. 64.0 g/eq b. 3 eq/mole
9. 3.81%
10. 26.3 g/eq
11. a. acidic b. 7.6 mL
12. a. 94.80 g/eq b. 37.9 g
    c. 0.0308 eq   d. 154 mL
13. a. 31.4 g b. 40.0 mL
14. a. 0.00275 eq b. 0.869 g
    c. 0.0158 d. 143 mL
15. 0.0660 N
16. 0.08250 N
17. 13.7% Fe
18. 71.5% (that is, 2.02 g of ZnO)
19. 225 mL

SPECIAL SECTION II
STUDY MATERIAL FOR
NORMALITY AND EQUIVALENTS

LEARNING GOALS:

1. Be able to determine the number of equivalents per mole and the equivalent mass for common acids and bases. From this information it should be possible to determine how many equivalents are contained in a given acid or base sample. (Sec. I.1)

2. Understand how to use the relaionship that states equivalents of acid = equivalents of base to solve problems involving neutralization reactions or acid-base titrations at the equivalence point. (Sec. I.2)

3. Given the reactants and products in a half reaction, be able to determine the number of equivalents per mole and the equivalent mass for oxidizing agents and reducing agents. Based on this information, be able to determine how many equivalents are contained in a sample of oxidizing or reducing agent. (Sec. I.3)

4. Using the equivalents of oxidizing agent = equivalents of reducing agent relationship, it's also necessary to be able to solve problems involving oxidation-reduction titrations at the equivalence point. (Sec. I.3)

5. For either acid-base or oxidation-reduction reactions, be able to calculate the new normality from an old normality, if given the chemical equations used to define these two quantities. (Sec. I.3)

IMPORTANT NEW TERMS:

chemical equivalent (acid-base reactions) (I.1)
chemical equivalent (redox reactions) (I.3)

end point (I.2)
equivalence point (I.2)
equivalent mass (I.1)
equivalents per mole (I.1)
normality (I.1)

CONCEPT TEST

1. For acid-base reactions, one equivalent of a substance is defined to be the amount of that substance that will combine with or produce one mole of _____.

2. For oxidation-reduction reactions, one equivalent of a substance is defined as the amount of that substance that will combine with or produce one mole of _____.

3. Indicate the number of equivalents per mole for each of the following acids and bases. (Assume complete neutralization.)

a. H_2SO_4 _____   b. NaOH _____

c. H_3PO_4 _____   d. $Ca(OH)_2$ _____

4. Indicate the number of equivalents per mole for the reactant underlined in each of the following balanced half reactions.

half reaction	eq/mol
a. <u>Zn</u> $\rightarrow$ Zn^{2+} + 2 e^-	_____
b. <u>Br_2</u> + 2 e^- $\rightarrow$ 2 Br^-	_____
c. <u>NO_3^-</u> + 4 H^+ + e^- $\rightarrow$ NO + 2 H_2O	_____

5. The _____ of a compound is the mass of that compound needed to produce one chemical equivalent.

6. _____ is defined as equivalents of solute per liter of solution.

7. The equivalence point of an acid-base titration occurs when the equivalents of acid equals the _____.

8. The _____ of an acid-base titration occurs when sufficient titrant has been added to cause a chemical indicator to change color.

STUDY HINTS:

1. As you will see, the titration problems discussed in this section are always at the equivalence point, which allows us to use the equivalents relationship. You may well wonder what to do if you find a problem in which the titration is not at the equivalence point. This is more complicated, and problems of this type will not be encountered until later in the course.

2. Be sure to understand the difference between equivalents and equivalents per mole. To draw a rough analogy, equivalents per mole is comparable to the cost of bread per loaf and equivalents can be compared with how much money you have. If bread costs 80 cents per loaf, that is a fixed amount, but doesn't determine how much money you have. It does mean that if you have $1.60, you can buy two loaves of bread. If a certain reactant is found to have two equivalents per mole, this doesn't mean you always must have two equivalents, but it does mean that if you have one mole you will have two equivalents. For a given chemical in a given reaction, the number of equivalents per mole is fixed, but you may have any number of equivalents depending on how many grams of substance are available.

3. Be sure to read the problems carefully to determine whether you are doing calculations that deal with an ion or a complete compound. For example, a problem can ask what is the percentage of iron(II) in a sample or what is the percentage of iron(II) chloride in a sample. It is easy to overlook this distinction in the heat of a test, but it makes a major difference in your answer.

PRACTICE PROBLEMS

1. a. Calculate the normality of a nitric acid solution prepared by dissolving 50.0 grams of HNO_3 in enough water to produce 235 ml of solution. You may assume that the nitric acid will be used in a simple acid-base reaction. b. What is the molarity of this solution? c. What is the concentration of H^+ in this solution?

2. Calculate the number of equivalents in a 125 gram sample of phosphoric acid, H_3PO_4. When phosphoric acid reacts with a strong base, such as NaOH, three moles of hydrogen ions can react per mole of acid.

3. How many milliliters of 0.110 N H_2SO_4 will be required to titrate 5.00 grams of KOH to the equivalence point?

4. What is the equivalent mass of a certain unknown acid if 27.3 ml of 0.210 N KOH is required to titrate a 0.361 gram sample of the unknown acid to the end point.

5. Calculate the normality of a solution of a sodium thiosulfate solution if 30.0 ml of this solution is required to titrate 1.43 grams of iodine according to the equation

$$I_2 \;+\; 2\,Na_2S_2O_3 \;\rightarrow\; 2\,NaI \;+\; Na_2S_4O_6$$

6. The skeleton equation (not balanced) for the reduction of iron(II) with permanganate in acidic solution is

$$Fe^{2+} \;+\; MnO_4^{-} \;\rightarrow\; Mn^{2+} \;+\; Fe^{3+}$$

Answer the following questions based on this equation.
(a) What is the equivalent mass of potassium permanganate?
(b) If you wish to produce a 0.200 N solution of potassium permanganate, how many grams of $KMnO_4$ must be dissolved in 2.00 liters of water?
(c) How many equivalents of iron(II) are present in a 5.00 gram sample of iron(II) chloride?
(d) How many milliliters of 0.200 N potassium permanganate solution will be required to exactly oxidize 5.00 grams of iron(II) chloride to iron(III) chloride?

7. The equivalent mass of benzoic acid is 122.1 g/eq. An unknown sample having a mass of 0.800 grams is a mixture of benzoic acid and inert material. If 24.7 ml of 0.206 N KOH is required to titrate this sample to the equivalence point, (a) how many grams of benzoic acid were in the original sample, and (b) what is the percent by weight of benzoic acid in the original sample?

PRACTICE PROBLEM SOLUTIONS

1. a. If we knew the equivalent mass of nitric acid, the normality could be calculated using the equation

$$\text{normality} = \frac{\text{g nitric acid/ eq mass nitric acid}}{\text{volume of solution (L)}}$$

In an acid-base reaction, there is one equivalent per mole for nitric acid, and so the equivalent mass is obtained by

$$\text{equivalent mass} = \frac{\text{molar mass of nitric acid}}{\text{eq/mol for nitric acid}}$$

$$\text{equivalent mass} = \frac{63.02 \text{ g/mol}}{1 \text{ eq/mol}}$$

$$\text{equivalent mass} = 63.02 \text{ g/eq}$$

94

Substituting into the equation for normality

$$\text{normality} = \frac{50.0 \text{ g} / 63.02 \text{ g/eq}}{0.235 \text{ L}}$$

normality = 3.38 N

b. Use the equation that relates normality and molarity

normality = molarity x eq/mol

$$\text{molarity} = \frac{\text{normality}}{\text{eq/mol}} = \frac{3.38 \text{ N}}{1 \text{ eq/mol}}$$

molarity = 3.38 M

c. The hydrogen ion concentration is numerically equal to the normality expressed in molar concentration units, and so

H^+ concentration = 3.38 M

2. The number of equivalents per mole is given as three in the problem. To find equivalents, we can use the equation

$$\text{equivalents} = \frac{\text{mass}}{\text{equivalent mass}}$$

The equivalent mass of H_3PO_4 is given by the equation

$$\text{equivalent mass} = \frac{\text{molar mass}}{\text{eq/mol}} = \frac{98.00 \text{ g/mol}}{3 \text{ eq/mol}}$$

equivalent mass = 32.67 g/eq

Substituting the equivalent mass into the equation for equivalents

$$\text{equivalents} = \frac{125 \text{ g}}{32.67 \text{ g/eq}}$$

equivalents = 3.83 eq

3. Since the reaction is at the equivalence point

equivalents of H_2SO_4 = equivalents of KOH

For sulfuric acid, we know the normality and wish to find the volume, so we can substitute

eq of H_2SO_4 = normality of H_2SO_4 x volume of H_2SO_4

For KOH, we know the mass, and can readily determine the equivalent mass, since there is 1 eq/mole for KOH.

$$\text{eq mass of KOH} = \frac{\text{molar mass}}{\text{eq/mol}} = \frac{56.11 \text{ g/mol}}{1 \text{ eq/mol}}$$

eq mass of KOH = 56.11 g/eq

$$\text{equivalents of KOH} = \frac{\text{grams of KOH}}{\text{equivalent mass of KOH}}$$

The equivalents relationship has now become

$$\text{normality } (H_2SO_4) \text{ x volume } (H_2SO_4) = \frac{\text{grams (KOH)}}{\text{eq mass (KOH)}}$$

Substituting the known values

$$0.110 \text{ N x V}(H_2SO_4) = \frac{5.00 \text{ g}}{56.11 \text{ g/eq}}$$

Solving for the volume of sulfuric acid (remember it will be in liters).

$V(H_2SO_4) = 0.810$ L

$V(H_2SO_4) = 0.810$ L x 1000 mL/L

$V(H_2SO_4) = \underline{8.10 \times 10^2 \text{ ml}}$

4. Since the titration is at the end point, begin with the relationship

equivalents of base = equivalents of acid

Examining the data suggests we should substitute N x V for equivalents of base and grams/equivalent mass for equivalents of the acid. The equivalents relationship now becomes

$$N(base) \times V(base) = \frac{grams\ (acid)}{equivalent\ mass\ (acid)}$$

Substitute the available information and solve for the equivalent mass. (Don't forget you must convert the volume to liters!)

$$0.210\ N \times 0.0273\ L = \frac{0.361\ g}{eq\ mass\ (acid)}$$

equivalent mass (acid) = 63.0 g/eq

5. If the iodine is to be completely titrated by the thiosulfate, this oxidation-reduction titration must be at the equivalence point, allowing us to write the relationship

eq (oxidizing agent) = eq (reducing agent)

Sodium thiosulfate is the reducing agent. We are given the volume and wish to find the normality of this reagent, so it's easy to see that we should substitute N x V for equivalents of reducing agent.

We are given the mass of iodine, and if we could find the equivalent mass, we could substitute the grams/equivalent mass = equivalents in the equivalents relationship. To find the equivalent mass of iodine, we must write the half reaction (notice that the NaI ionizes to form iodide ions).

$$2e + I_2 \rightarrow 2\ I^-$$

Which indicates that there are 2 eq/mole for I_2.

Determine the equivalent mass using the equation

$$eq\ mass = \frac{molar\ mass}{eq/mole} = \frac{253.8\ g/mol}{2\ eq/mol} = 126.9\ g/eq$$

Now the specific equivalents relationship for this problem is

$$\frac{\text{grams } (I_2)}{\text{eq mass } (I_2)} = N\ (S_2O_3{}^{2-}) \times V\ (S_2O_3{}^{2-})$$

Now substitute the values given (but be sure to change the volume to liters) and solve.

$$\frac{1.43\ g}{126.9\ g/eq} = N\ (S_2O_3{}^{2-}) \times 0.0300\ L$$

$$\underline{N\ (S_2O_3{}^{2-}) = 0.376\ N}$$

6. a. First, write and balance the permanganate half reaction to determine the equivalents per mole

$$8H^+ + 5e + MnO_4^- \rightarrow Mn^{2+} + 4\ H_2O$$

There are 5 eq/mol for permanganate.
Next, determine the equivalent mass

$$\text{eq mass} = \frac{\text{molar mass}}{\text{eq/mol}} = \frac{158.04\ g/mol}{5\ eq/mol}$$

$$\underline{\text{eq mass} = 31.61\ g/eq}$$

Since the mass of potassium permanganate is indicated, use the molar mass of $KMnO_4$, not just the MnO_4^- ion.

b. Solve the equation for the number of equivalents

$$\text{eq} = N \times V = 0.200\ N \times 2.00\ L = 0.400\ eq$$

and then multiply this number of equivalents times the equivalent mass to determine the number of grams needed.

$$\text{grams} = \text{eq} \times \text{eq mass} = 31.61\ g/eq \times 0.400\ eq$$

$$\underline{\text{grams} = 12.6\ g\ KMnO_4}$$

c. From the information given, we can write and balance the half reaction for iron

98

$$Fe^{2+} \rightarrow Fe^{3+} + e$$

which tells us that there is one equivalent per mole for iron(II) in this reaction.

Next, find the equivalent mass for iron(II) chloride (not just for the iron!).

$$\text{eq mass } (FeCl_2) = \frac{\text{molar mass}}{\text{eq/mole}} = \frac{126.7 \text{ g/mol}}{1 \text{ eq/mol}}$$

$$\text{equivalent mass } (FeCl_2) = 126.7 \text{ g/eq}$$

To find the number of equivalents, divide the grams by the equivalent mass

$$\text{equivalents} = \frac{5.00 \text{ g}}{126.75 \text{ g/eq}}$$

equivalents = 0.0394 eq

d. This titration is at the equivalence point, and so we may write

equivalents of MnO_4^- = equivalents of $FeCl_2$

We can substitute N x V for equivalents of permanganate, and the equivalents of $FeCl_2$ were obtained in the previous section.

N (MnO_4^-) x V (MnO_4^-) = equivalents of $FeCl_2$

Substituting the values given

0.200 N x V (MnO_4^-) = 0.0394 eq

V (MnO_4^-) = 0.197 L

but the volume is requested in milliliters

V (MnO_4^-) = 0.197 L x 1000 mL/L

V (MnO_4^-) = 197 ml

7. a. As usual, begin with the equivalents relationship

equivalents of KOH = equivalents of benzoic acid

Substitute V x N for equivalents of KOH and grams/equivalent mass for equivalents of benzoic acid. Notice that the mass given is <u>not</u> the mass of pure benzoic acid!

$$\text{V (KOH) x N (KOH)} = \frac{\underline{\text{grams (benzoic acid)}}}{\text{equivalent mass (benzoic acid)}}$$

$$\text{0.0247 L x 0.206 N} = \frac{\underline{\text{grams (benzoic acid)}}}{\text{122.1 g/eq}}$$

<u>grams (benzoic acid) = 0.621 g</u>

b. Now find the percent of the sample that is benzoic acid

$$\text{percent acid} = \frac{\underline{\text{grams of acid x 100}}}{\text{grams of sample}}$$

$$\text{percent acid} = \frac{\underline{\text{0.621 g x 100}}}{\text{0.800 g}}$$

<u>percent acid = 77.6 %</u>

PRACTICE TEST (40 Minutes)

1. Indicate the number of equivalents per mole for each of the following acids and bases. (Assume complete neutralization.)

a. H_2CO_3 _____   b. HNO_3 _____

c. NH_3 _____   d. $Ba(OH)_2$ _____

2. In each case on the next page, the reactant and product are indicated for a skeleton oxidation-reduction half reaction. In each case, write the balanced half reaction (in acidic solution) and indicate the equivalents per mole for the reactant.

| half reaction | eq/mol |

a. Aluminum(III) is converted to _____
   aluminum metal.

b. Dichromate ion ($Cr_2O_7^{2-}$) is _____
   converted to chromium(III).

c. Oxalate ion ($C_2O_4^{2-}$) is _____
   is converted to carbon dioxide.

3. (a) How many equivalents of base are contained in 250.0 ml of 0.231 N calcium hydroxide, $Ca(OH)_2$? (b) If you dilute this solution to a volume of 500. ml, what is the new normality? (c) Did the number of equivalents of base change when the solution was diluted?

4. Malic acid, $C_4H_6O_5$, which is commonly found in apples and other fruits, normally loses two hydrogen ions on reaction with a base. (a) What is the equivalent mass of this acid? (b) If 1.75 grams of malic acid is used in a reaction, how many equivalents of acid are used? (c) If the 1.75 gram sample of malic acid is dissolved in enough water to make 150. ml of solution, what is the normality and molarity of the solution?

5. A certain chromate solution (CrO_4^{2-}) is 3.0 N when it forms chromium(III) hydroxide in basic solution. Write the balanced half reaction and determine the molarity of the chromate solution.

6. (a) How many milliliters of 0.250 N KOH would be required to neutralize completely 2.50 grams of oxalic acid, $H_2C_2O_4$? (Oxalic acid has two ionizable hydrogens.) (b) How many milliliters of 0.250 N $Ba(OH)_2$ would be require to neutralize the same mass of oxalic acid?

7. What is the normality of a Sn^{2+} if it contains 1.18 grams of tin(II) per liter of solution and the tin will be used as a reducing agent according to the half reaction

$$Sn^{2+} \rightarrow Sn^{4+} + 2e^-$$

8. A sample of iron ore weighing 0.800 grams is dissolved in acid and all of the iron in the sample is reduced to iron(II). The sample is then titrated in acid solution with 0.100 Normal $KMnO_4$ according to the unbalanced equation

$$Fe^{2+} + MnO_4^- \rightarrow Mn^{2+} + Fe^{3+}$$

If exactly 28.0 ml of potassium permanganate is required to reach the end point of this titration, what is the percentage of iron in the original sample?

CONCEPT TEST ANSWERS

1. hydrogen ions (or protons)   2. electrons
3. a. two b. one c. three d. two
4. a. two b. two c. one
5. equivalent mass   6. normality
7. equivalents of base   8. end point

PRACTICE TEST ANSWERS

1. a. two   b. one   c. one   d. two
2.   <u>half reaction</u>   <u>eq/mol</u>

a. $3 e^- + Al^{3+} \rightarrow Al$ 3

b. $6 e^- + 14 H^+ + Cr_2O_7^{2-}$
$\rightarrow 2 Cr^{3+} + 7 H_2O$ 6

c. $C_2O_4^{2-} \rightarrow 2 CO_2 + 2e^-$ 2

3. a. 0.0578 equivalents b. 0.116 c. no
4. a. 67.1 g/eq b. 0.0261 eq
   c. 0.174 N, 0.0870 M
5. $3e^- + 4 H_2O + CrO_4^{2-} \rightarrow Cr(OH)_3 + 5 OH^-$

   Molarity = 1.0 M
6. a. 222 ml b. 222 ml
   (Notice the amounts <u>must be the same</u>!)
7. 0.0199 N   8. 19.5%

CHAPTER 6
THERMOCHEMISTRY

LEARNING GOALS:

1. Recognize the various forms of energy and understand how the transformations from one form to another are governed by the conservation of energy principle. Also be able to convert energy values from one unit to another, for example, from calories to joules and from joules to calories. (Sec. 6.1 & 6.2)

2. Adding or removing heat causes a substance to change temperature and/or state. If no state change is involved, the specific heat equation (or the molar heat capacity equation) is used.

$$q = c \times m \times \Delta t$$

The equation for state changes is

$$q = \Delta H \times m$$

where ΔH may be the value for fusion or vaporization. Be able to do problems that use either (or both) of these relationships. Your instructor may also ask you to memorize the values of the heat of vaporization and fusion for water. (Sec. 6.3)

3. Understand the first law of thermodynamics, and terms related to this concept, including endothermic and exothermic reactions, energy change, and enthalpy change. (Sec. 6.4)

4. If given an appropriate set of equations with associated enthalpy values, be able to use Hess's Law to calculate the enthalpy change for a reaction that is a simple combination of the given equations. (Sec. 6.5)

5. Know the special properties of state functions

and recognize which of the thermodynamic variables discussed so far are state functions. (Sec. 6.6)

6. Understand the meaning of the term standard conditions, be able to predict the standard states for common substances, and be able to use the equation

$$\Delta H^{\circ}_{rxn} = \Sigma \; \Delta H^{\circ}_f \; (products) - \Sigma \; \Delta H^{\circ}_f \; (reactants)$$

to calculate the standard enthalpy change for reactions based on values from an appropriate table of standard enthalpies of formation. (Sec. 6.7)

7. Understand the basic experimental procedures of calorimetry, that is, the method used to measure heats of reactions in the laboratory. (Sec. 6.8)

IMPORTANT NEW TERMS

calorie (6.2)
calorimeter (6.8)
calorimetry (6.8)
electrical energy (6.1)
endothermic process (6.4)
energy (6.1)
enthalpy of formation (6.7)
enthalpy of combustion (6.5)
enthalpy change (6.4)
exothermic process (6.4)
first law of thermodynamics (6.4)
heat (thermal energy) (6.1)
heat capacity (6.3)
Hess's law of constant heat summation (6.5)
joule (6.2)
kilocalorie (6.2)
kinetic energy (6.1)
latent heat of vaporization (6.3)
latent heat of fusion (6.3)
molar heat capacity (6.2)
potential energy (6.1)
principle of conservation of energy (6.1)
quantity of heat (6.3)
radiant energy (6.1)
specific heat (6.3)
standard conditions (6.7)
standard state (6.7)

CONCEPT TEST

1. The science of heat or energy flow in chemical reactions is called _____.

2. The amount of heat required to raise the temperature of 1.00 gram of pure liquid water from $14.5^{\circ}C$ to $15.5^{\circ}C$ is called a(n) _____.

3. _____ is the name of the SI unit of energy; it is equivalent to _____ calories.

4. The amount of heat required to raise the temperature of any specific substance by one degree Celsius is called the _____ of that substance.

5. The specific heat of aluminum is 0.902 J/g·K; what is the molar heat capacity of aluminum?

_____

6. The material being studied in a thermodynamic experiment is called the _____, and the remainder of the universe is called the

_____.

7. The standard enthalpy of formation for any

element is _____ under standard conditions.

8.   State the First Law of Thermodynamics and define each symbol in the equation.

_____

_____

9. The value of a(n) _____ depends only on the state of a system; therefore, the change in value of such a function is determined only by its initial and final value.

10. The standard state of an element or compound is measured at a pressure of _____ and a temperature of _____.

STUDY HINTS:

1. The properties of state functions are the key to most of the problem-solving methods described in this chapter; therefore, it's critical that you understand the meaning of the term, state function.

2. Thermodynamic symbols often include subscripts and superscripts that represent critical information about the conditions, such as temperature and pressure.   For example, the superscript o on the symbol ΔH^o should tell you the temperature, the pressure, and the physical state of the substance involved.   Don't overlook this valuable source of information.

3. Some students have difficultly distinguishing the different problem types in this chapter.   It may be helpful to note that learning goal 2 usually refers to systems undergoing physical change, that is, melting, freezing, or temperature change. Learning goals 4 and 6 describe closely related problems, but in the first case the data is usually

a list of equations with the corresponding enthalpy values and in the second case the data is provided as a table of enthalpy values. Try to identify these three different situations as you look at the problems in the text and in this guide.

4. Be sure that all of the equations are balanced, before you begin to do a thermochemistry problem. Failure to do this can waste time (if you notice the error later on) or cause your solution to be incorrect.

PRACTICE PROBLEMS

1. (L1) The box of a popular breakfast cereal states that one serving provides 50 Calories (without milk). How much energy is this if measured in joules? (Hint; Notice that there is a capital C on the energy unit, so it's really a kilocalorie!)

2. (L2) How many joules of energy are required to change 100.0 grams of ice at $0.0°C$ to water at $40.0°C$. The heat of fusion for water is 333 J/g.

3. (**L6**) Calculate the Standard Enthalpy of Combustion for the reaction

$C_2H_4(g)$ + $O_2(g)$ $\rightarrow$ $CO_2(g)$ + $H_2O(g)$

Based on the following Standard Enthalpies of Formation: $C_2H_4(g)$ +52.3 kJ/mol; $CO_2(g)$ -393.5 kJ/mol; and $H_2O(g)$ -241.8 kJ/mol

4. (**L4**) Cyanamide, CH_2N_2, is a weak acid that is sometimes used as a fertilizer. Calculate the standard enthalpy of formation for cyanamide, given the following standard enthalpies of reaction:

$CH_2N_2(s)$ + 3/2 $O_2(g)$
$\rightarrow$ $CO_2(g)$ + $H_2O(\ell)$ + $N_2(g)$ ΔH_1= -741.4 kJ/mol

$C(s)$ + $O_2(g)$ $\rightarrow$ $CO_2(g)$ ΔH_2= -393.5

$H_2(g)$ + 1/2 $O_2(g)$ $\rightarrow$ $H_2O(\ell)$ ΔH_3= -285.8

5. (**L6**) Phosgene, $POCl_2(g)$, and hydrogen sulfide are both very poisonous, but when they are heated together the product is carbon disulfide, an evil-smelling gas that is much less toxic than either of the reactants. Calculate the standard enthalpy of formation for phosgene given that the standard enthalpy of reaction is -49.0 kJ for

$$COCl_2(g) + 2\ H_2S(g) \rightarrow 2\ HCl(g) + H_2O(g) + CS_2(g)$$

and using the following standard enthalpies of formation:

	ΔH_f^o		ΔH_f^o
$CS_2(g)$	117.4 kJ/mol	$H_2S(g)$	-20.63 kJ/mol
$HCl(g)$	- 92.3	$H_2O(g)$	-241.8

6. (**L7**) If 1.15 grams of glucose (molar mass = 180.2 g/mol) is burned in a calorimeter, the temperature increases from $23.40^\circ C$ to $27.21^\circ C$. If the heat capacity of the calorimeter is 958 j/$^\circ C$ and the calorimeter contains 899.5 grams of water, calculate the amount of heat given off per mole of glucose under these conditions.

7. (**L4**) Coke reacts with steam to produce a mixture called coal gas, which can be used as a fuel or as a starting material for other reactions. The equation for the production of coal gas is

$$2 \; C(s) \; + \; 2 \; H_2O(g) \; \rightarrow \; CH_4(g) \; + \; CO_2(g)$$

Determine the Standard Enthalpy Change for this reaction based on the following Standard Enthalpies of Reaction:

$$C(s) \; + \; H_2O(g) \; \rightarrow \; CO(g) \; + \; H_2(g) \qquad \Delta H^\circ = +131.3 \text{ kJ}$$

$$CO(g) \; + \; H_2O(g) \; \rightarrow \; CO_2(g) \; + \; H_2(g) \qquad \Delta H^\circ = -41.2$$

$$CH_4(g) \; + \; H_2O(g) \; \rightarrow \; 3 \; H_2(g) \; + \; CO(g) \qquad \Delta H^\circ = +206.1$$

PRACTICE PROBLEM SOLUTIONS

1. Remember that the calories used in this case are one thousand times as large as the calorie units we normally use and one of these "small" calories is 4.184 joules. Thus the conversion is

joules = 50 Cal x 1000 cal/Cal x 4.184 J/cal
joules = 200,000 joules

2. First, calculate how much heat is needed to melt the ice at the normal melting point, 0°C.

$$q = \Delta H_{fus} \; x \; m \; = \; 333 \text{ J/g x } 100.0 \text{ g} \; = 33300 \text{ J}$$

Next, determine how much heat is required to heat the liquid water 40 degrees Celsius.

$$q = c \times m \times \Delta t$$
$$= 4.184 \text{ J/g}^{\circ}\text{C} \times 100.0 \text{ g} \times 40.0^{\circ}\text{C}$$
$$q = 16736 \text{ J}$$

Finally, add heat needed for the two processes and round to the correct number of significant figures.

$$q = 33300 \text{ J} + 16736 \text{ J}$$
$$\underline{q = 50,000 \text{ J}}$$

3. First, balance the equation

$$C_2H_4(g) + 3 O_2(g) \rightarrow 2 CO_2(g) + 2 H_2O(g)$$

Now combine the standard enthalpy of formation values provided:

$$\Delta H^{\circ}_{rxn} = \Sigma \Delta H^{\circ}_f \text{ (products)} - \Sigma \Delta H^{\circ}_f \text{ (reactants)}$$

$$\Delta H^{\circ} = 2 \Delta H^{\circ}_f [CO_2(g)] + 2 \Delta H^{\circ}_f [H_2O(g)] - \Delta H^{\circ}_f [C_2H_4(g)]$$

$$\Delta H^{\circ} = 2(-393.5 \text{ kJ/mol})$$
$$+ 2(-241.8 \text{ kJ/mol}) - (52.3 \text{ kJ/mol})$$

$$\underline{\Delta H^{\circ} = -1322.9 \text{ kJ}}$$

4. The equation for the formation of cyanamide from the elements in their standard states is

$$C(s) + H_2(g) + N_2(g) \rightarrow CH_2N_2(s)$$

To obtain this equation, reverse the first equation

$$CO_2(g) + H_2O(\ell) + N_2(g) \rightarrow CH_2N_2(s) + 3/2 O_2(g)$$

Use the second equation as written

$$C(s) + O_2(g) \rightarrow CO_2(g)$$

And use the third equation as written

$$H_2(g) + 1/2 O_2(g) \rightarrow H_2O(\ell)$$

Adding these three equations produces the desired reaction, thus the enthalpy change for this reaction must be equal to

$$\Delta H = -\Delta H_1 + \Delta H_2 + \Delta H_3$$

$$\Delta H = -(-741.4 \text{ kJ}) + (-393.5 \text{ kJ}) + (-285.8 \text{ kJ})$$

$$\underline{\Delta H = +62.1 \text{ kJ/mol}}$$

5. The equation is already balanced, so substitute into the Hess equation:

$$\Delta H^{\circ}_{rxn} = \Sigma \Delta H^{\circ}_f \text{ (products)} - \Sigma \Delta H^{\circ}_f \text{ (reactants)}$$

$$\Delta H^{\circ} = 2 \Delta H_f^{\circ} [HCl(g)] + 2 \Delta H_f^{\circ} [H_2O(g)] - \Delta H_f^{\circ} [CS_2(g)]$$

$$+ 2 \Delta H_f^{\circ} [COCl_2(g)] - 2 \Delta H_f^{\circ} [H_2S(g)]$$

$$-49.0 \text{ kJ} = 2(-92.3 \text{kJ/mol}) + (-241.8 \text{ kJ/mol})$$

$$+ (117.4 \text{ kJ/mol}) - \Delta H_f^{\circ} [COCl_2(g)]$$

$$- 2(-20.63 \text{ kJ/mol})$$

$$\Delta H_f^{\circ} [COCl_2(g)] = \underline{218.7 \text{ kJ/mol}}$$

6. The heat absorbed by the bomb is given by

$$q_{bomb} = (C_{bomb})(\Delta t) = 958 \text{ J/}^{\circ}\text{C} \times (27.21-23.40)$$

$$= 3.65 \times 10^3 \text{ J}$$

The heat absorbed by the water equals

$$q_{water} = (4.184 \text{ J/g} \cdot {}^{\circ}\text{C})(m_{water})(\Delta t)$$

$$= (4.184 \text{ J/g} \cdot {}^{\circ}\text{C})(899.5 \text{ g})(27.32-23.40)$$

$$= 14.3 \times 10^3 \text{ J}$$

To find the heat generated by the combustion, add these two heat values and reverse the sign.

Total heat by 1.15 g of glucose = -18.0×10^3 J

$$= -18.0 \text{ kJ}$$

Now find the heat evolved per gram of glucose and multiply by the molar mass of this compound.

$$\Delta H = \frac{(-18.0 \text{ kJ})(180.2 \text{ g/mol})}{1.15 \text{ g}}$$

$$\underline{\Delta H = -2820 \text{ kJ/mol}}$$

7. To obtain the equation specified, double the first reaction

$$2 \times [C(s) + H_2O(g) \rightarrow CO(g) + H_2(g)]$$

use the second reaction as written

$$CO(g) + H_2O(g) \rightarrow CO_2(g) + H_2(g)$$

and reverse the third reaction.

$$3 H_2(g) + CO(g) \rightarrow CH_4(g) + H_2O(g)$$

Adding these three reactions and cancelling will produce

$$2 C(s) + 2 H_2O(g) \rightarrow CH_4(g) + CO_2(g)$$

and so the standard enthalpy of this reaction must equal

$$\Delta H = 2 \times (\Delta H_1) + \Delta H_2 - \Delta H_3$$

$$\Delta H = 2 \times (+131.3 \text{kJ}) + (-41.2 \text{ kJ}) - (+206.1 \text{ kJ})$$

$$\underline{\Delta H = +15.3 \text{ kJ}}$$

PRACTICE TEST (45 Minutes)

1. What is the standard state of each of the following elements at $25°C$ and 1 atmosphere pressure. Make your answer as specific as possible, for example, write $O_2(g)$, not just O_2.
a. carbon
b. fluorine
c. carbon dioxide
d. water

2. Calculate the standard enthalpy change for the reaction

$$2\ NO_{2(g)} \quad + \quad 7\ H_{2(g)} \quad \rightarrow \quad 2\ NH_{3(g)} \quad + \quad 4\ H_2O_{(g)}$$

Based on the following standard enthalpies of formation: $NO_2(g)$ +33.2 kJ/mol; $H_2O(g)$ -242 kJ/mol; and $NH_3(g)$ -46.1 kJ/mol

3. Write the equation for the formation of hydrazine gas, $N_2H_4(g)$, from the elements under standard conditions and then calculate the Standard Enthalpy of Formation for this compound, given the standard enthalpies of reaction for the following equations:

$$N_2H_4(g) \quad + \quad O_2(g)$$
$$\rightarrow \quad N_2(g) \quad + \quad 2\ H_2O(g) \qquad \Delta H^\circ = -579\ kJ$$

$$H_2(g) \quad + \ 1/2\ O_2(g) \quad \rightarrow \quad H_2O(g) \qquad \Delta H^\circ = -242$$

4. Calculate the number of joules necessary to heat 58.3 grams of solid mercury from its freezing point ($-39^\circ C$) to a temperature of $100^\circ C$. For liquid mercury the specific heat is 1.26 J/g$\cdot^\circ C$ and the heat of fusion is 11.6 J/g.

5. The combustion of acetylene, C_2H_2, is used to produce high temperatures for welding torches. Calculate the Standard Enthalpy of Reaction for the balanced equation

$$2\ C_2H_2(g) \quad + \quad 5\ O_2(g) \quad \rightarrow \quad 4\ CO_2(g) \quad + \quad 2\ H_2O(\ell)$$

Given the Standard Enthalpies of Reaction for the following:

$$2\ C(graphite) + \quad H_2(g) \qquad \rightarrow C_2H_2(g) \qquad \Delta H_1 = \quad 227\ kJ$$

$$C(graphite) \quad + \quad O_2(g) \qquad \rightarrow CO_2(g) \qquad \Delta H_2 = -394$$

$$H_2(g) \qquad \quad + \ 1/2\ O_2(g) \rightarrow H_2O(\ell) \qquad \Delta H_3 = -286$$

6. Write the equation for the formation of $H_2O_2(g)$, gaseous hydrogen peroxide, from the elements at $25^\circ C$ in their standard states, and then calculate the

standard enthalpy of formation for this compound based on the following reactions:

$$2 \text{ H(g)} + 2 \text{ O(g)} \rightarrow \text{H}_2\text{O}_2\text{(g)} \quad \Delta H^{\circ} = -1071 \text{ kJ}$$

$$\text{O}_2\text{(g)} \rightarrow 2 \text{ O(g)} \quad \Delta H^{\circ} = 498.3$$

$$2 \text{ H(g)} + \text{O(g)} \rightarrow \text{H}_2\text{O(g)} \quad \Delta H^{\circ} = -926.9$$

$$\text{H}_2\text{(g)} + 1/2 \text{ O}_2\text{(g)} \rightarrow \text{H}_2\text{O(g)} \quad \Delta H^{\circ} = -241.8$$

7. When 0.500 grams of sucrose (molar mass = 342.3g/mol) is burned in a calorimeter, the temperature increases from 21.11°C to 23.42°C. Calculate the heat released per mole of sucrose, if the heat capacity of the calorimeter is 0.852 kJ/$^{\circ}$C and the calorimeter contains 650.0 grams of water.

8. One way to conserve gasoline might be by marketing a blend of ethanol and gasoline as an automobile fuel. Write the equation for the combustion of ethanol, $\text{C}_2\text{H}_5\text{OH(g)}$, and use the standard enthalpies of formation given below to calculate the standard enthalpy of combustion per mole of ethanol.

Standard Enthalpies of Formation

$\text{C}_2\text{H}_5\text{OH(g)}$ −235.1 kJ/mol $\text{CO}_2\text{(g)}$ −393.5 kJ/mol
$\text{H}_2\text{O}(\ell)$ −285.8

9. The organic compound ethylene, C_2H_4, reacts with hydrogen to form ethane, C_2H_6:

$$\text{H}_2\text{C=CH}_2\text{(g)} + \text{H}_2\text{(g)} \rightarrow \text{H}_3\text{C-CH}_3\text{(g)}$$

Calculate the enthalpy of reaction for this reaction based on the following:

$$\text{H}_2\text{C=CH}_2\text{(g)} + 3 \text{ O}_2\text{(g)}$$
$$\rightarrow 2 \text{ CO}_2\text{(g)} + 2 \text{ H}_2\text{O(g)} \quad \Delta H^{\circ} = -1323 \text{ kJ}$$

$$2 \text{ H}_2\text{(g)} + \text{O}_2\text{(g)} \rightarrow 2 \text{ H}_2\text{O(g)} \quad \Delta H^{\circ} = -483.6$$

$$\text{H}_3\text{C-CH}_3\text{(g)} + 7/2 \text{ O}_2\text{(g)}$$
$$\rightarrow 2 \text{ CO}_2\text{(g)} + 3 \text{ H}_2\text{O(g)} \quad \Delta H^{\circ} = -1428$$

CONCEPT TEST ANSWERS

1. thermodynamics
2. calorie
3. joule, 0.239 cal/joule
4. specific heat
5. 24.4 J/mol·K
6. system, surroundings
7. zero
8. $\Delta E + w = q_p$, ΔE is the change in internal energy, w is the work done by the system, and q_p is the heat gained or lost by the system at constant pressure.
9. state function
10. 1 atm, 25°C

PRACTICE TEST ANSWERS

1. (**L6**) a. C(graphite) b. $F_2(g)$

   c. $CO_2(g)$ d. $H_2O(\ell)$
2. (**L6**) -1127 kilojoules
3. (**L4**) $N_2(g) + 2 H_2(g) \rightarrow N_2H_4(g)$

   +95 kilojoules
4. (**L2**) 10,900 joules
5. (**L4**) -2602 kJ
6. (**L4**) $H_2(g) + O_2(g) \rightarrow H_2O_2(g)$

   -137 kJ
7. (**L7**) -5650 kJ/mol
8. (**L4**) $C_2H_5OH(g) + 3 O_2(g) \rightarrow 2 CO_2(g) + 3 H_2O(g)$

   -1409.3 kJ/mol
9. (**L4**) -137 kJ

CHAPTER 7
NUCLEAR CHEMISTRY

LEARNING GOALS:

1. Be able to write nuclear reactions for various types of radioactive processes, including nuclear fission, nuclear fusion, spontaneous radioactive decay, and artificial transmutations. (Sec. 7.2)

2. Be familiar with the factors that are important in determining nuclear stability and understand the importance of binding energy in the determination of nuclear stability. (Sec. 7.3)

3. Be able to do calculations involving the relationship between elapsed time and the amount of radioactive material which has been transformed, using the equation

$$\ln \frac{N}{N_o} = -kt$$

This equation is also the basis for radiochemical dating calculations. (Sec. 7.4)

4. Nuclear fission reactors are currently used to generate a significant fraction of the world's electric power and nuclear fusion is hoped to eventually play a similar role. Understand the basic principles that govern these types of power reactors as well as the assets and liabilities that these technologies present. (Sec. 7.6 & 7.7)

5. Be familiar with some of the commercial applications of radiation in medicine, chemical research, and chemical analysis. (Sec. 7.9)

IMPORTANT NEW TERMS:

activity (7.4)
alpha radiation (7.1)
artificial transmutations (7.5)
beta particles (7.1)
binding energy (7.3)

chain reaction (7.6)
Curie (7.1)
decay constant (7.4)
gamma radiation (7.1)
half-life (7.4)
K-capture (7.2)
magic numbers (7.3)
neutron activation analysis (7.9)
nuclear reactor (7.7)
nuclear fission (7.6)
nuclear fusion (7.8)
nucleons (7.2)
positron (7.2)
rad (7.1)
radioactive series (7.2)
radioactivity (7.1)
radiochemical dating (7.4)
radioisotope tracers (7.9)
rem (7.1)
Röntgen (7.1)
transuranium elements (7.5)

CONCEPT TEST

1. Alpha radiation consists of _____,
and beta radiation consists of _____.
2. _____ radiation has the greatest
ability to penetrate matter.
3. In a correct equation for a nuclear reaction,
the sum of the mass numbers of the reacting nuclei
must equal _____,
and the sum of the atomic numbers of the products
must equal the sum of _____.
4. All isotopes having an atomic number greater
than that of the element _____ are
radioactive.
5. List the four stable isotopes that have both an
odd number of protons as well as an odd number of
neutrons. _____

6. The point of maximum stability in the binding energy curve occurs in the vicinity of what isotope of what element? _____

7. The _____ of a radioactive sample is a measure of the rate of nuclear decay and can be measured using a Geiger counter.

8. Radiochemical dating is based on the assumption that the _____ of each radioactive isotope was the same in ancient times as it is now.

9. The energy required to separate a nucleus into its individual nucleons (or the energy evolved when these nucleons combine to form a nucleus) is called the _____.

10. If a nuclear fission reaction produces more neutrons than would be necessary to cause another fission reaction to occur, the result can be a sequence of rapidly occurring reactions called a(n) _____.

11. The rate of the fission in a nuclear reactor is controlled by inserting rods made of substances such as _____ that will strongly absorb neutrons. These control rods should not be confused with the moderator, substances such as water or graphite, that slow down the neutrons without absorbing them.

12. Approximately _____ % of the electric power in the United States is produced by nuclear reactors.

13. Nuclear fusion reactions of the element _____ are the primary source of energy from the sun.

14. When a sample is exposed to neutrons, each constituent element forms a specific radioisotope that can be identified by its characteristic radiation and half-life. This observation is the basis of the non-destructive method of elemental analysis called _____.

15. Stable isotopes usually have an _____ (choose from odd or even) number of protons and an _____ (choose from odd or even) number of neutrons.

STUDY HINTS:

1. Be sure that you can identify the common types of nuclear radiation, that is, alpha, beta, and gamma (α, β, and γ) by name, by appropriate symbol, and in terms ofthe nature of the radiation. Questions on nuclear reactions may list the type of radiation emitted either by sumbol or by description.

2. Your instructor may wish you to know some of the terminology used to describe radioactive exposure, such as Curie, Röntgen, rad, rem, fatal dose, and background radiation. Listen carefully to determine if this is required.

3. You will probably be using a calculator to do the problems on radioactive decay. Notice that there are two different types of logarithms on your calculator, base 10 logs (usually designated **log**) and natural logs (designated **ln x**). You will use the latter type of logs for these problems; try to avoid confusing the two sets of calculator keys. This is also a good time to make sure that you know how to obtain an antilog using your calculator.

4. You must read radioactive decay problems carefully. The quantity, N, in the equation is the amount of material <u>remaining</u>. Sometimes a problem will refer to the amount of matter that has reacted. Be aware of this possible source of confusion as you are workng problems, and watch for this situation on examinations.

1. (**L1**) Fill in the single missing species to complete each reaction.

a. $^{234}_{90}Th \rightarrow$ ____ $+ ^{230}_{88}Ra$

b. $^{11}_{5}B + ^{4}_{2}He \rightarrow ^{1}_{0}n +$ _____

c. ____ $\rightarrow ^{0}_{-1}\beta + ^{127}_{53}I$

d. $^{131}_{56}Ba + ^{0}_{-1}\beta \rightarrow$ _____

e. $^{27}_{13}Al + ^{3}_{1}H \rightarrow$ _____ $+ ^{27}_{12}Mg$

f. $^{2}_{1}H +$ ____ $\rightarrow ^{4}_{2}He + ^{1}_{0}n$

g. $^{40}_{19}K \rightarrow ^{40}_{18}Ar +$ _____

h. $^{55}_{26}Fe +$ ____ $\rightarrow ^{55}_{25}Mn$

2. (**L2**) Each of these pairs of nuclides consists of one stable isotope and one that is unstable. Select the isotope that is more likely to be stable in each case and explain your choice.

a. $^{96}_{43}Tc$ or $^{96}_{42}Mo$ b. $^{14}_{7}N$ or $^{30}_{15}P$

3. (**L3**) Underwater archaeologists discovered a sunken ship off the coast of Israel just south of Haifa. The ship's cargo included an amazingly well-preserved cargo of lead and copper ingots and is thought to have sunk during the Late Bronze Age (1550-1200 B.C.). Radiocarbon dating on wood from this ship shows a carbon-14/carbon-12 ratio that is 0.673 times that for comparable wood from the

present time. If the half-life of carbon-14 is 5730 years, does the observed radiocarbon date agree with the suspected age of the ship?

4. (**L1**) a. One of the well-known radioactive decay series begins with uranium-238 and proceeds by emission of the following sequence of particles: alpha, beta, beta, alpha, alpha, alpha, and alpha. What is the isotope formed at conclusion of this set of processes?
b. Another well-known radioactive decay series begins with thorium-232 and proceeds by emission of the following sequence of particles: alpha, beta, beta, alpha, alpha, alpha, and alpha. What is the isotope formed at the conclusion of this set of processes?

5. (**L2**) Calculate the binding energy in kilojoules per mole of boron for the formation of boron-10. You will need to know the following masses:
$_{1}^{1}H$ = 1.00783; $_{0}^{1}n$ = 1.00867; and $_{5}^{10}B$ = 10.01294

PRACTICE PROBLEM SOLUTIONS

1. a. mass number: 234 = A + 230, thus A = 4
   atomic number: 90 = Z + 88, thus Z = 2
   the missing species is helium.

$$^{234}_{90}\text{Th} \rightarrow \ ^{4}_{2}\text{He} + \ ^{230}_{88}\text{Ra}$$

   b. mass number: 11 + 4 = 1 + A, thus A = 14
   atomic number: 5 + 2 = 0 + Z, thus Z = 7
   the missing species is nitrogen.

$$^{11}_{5}\text{B} + \ ^{4}_{2}\text{He} \rightarrow \ ^{1}_{0}\text{n} + \ ^{14}_{7}\text{N}$$

   c. mass number: A = 0 + 127, thus A = 127
   atomic number: Z = -1 + 53, thus Z = 52
   the missing species is iodine.

$$^{127}_{52}\text{Te} \rightarrow \ ^{0}_{-1}\beta + \ ^{127}_{53}\text{I}$$

   d. mass number: 131 + 0 = A, thus A = 131
   atomic number: 56 - 1 = Z, thus Z = 55
   the missing species is cesium.

$$^{131}_{56}\text{Ba} + \ ^{0}_{-1}\beta \rightarrow \ ^{131}_{55}\text{Cs}$$

   e. mass number: 27 + 3 = A + 27, thus A = 3
   atomic number: 13 + 1 = Z + 12, thus Z = 2
   the missing species is helium.

$$^{27}_{13}\text{Al} + \ ^{3}_{1}\text{H} \rightarrow \ ^{3}_{2}\text{He} + \ ^{27}_{12}\text{Mg}$$

   f. mass number: 2 + A = 4 + 1, thus A = 3
   atomic number: 1 + Z = 2 + 0, thus Z = 1
   the missing species is hydrogen.

$$^{2}_{1}\text{H} + \ ^{3}_{1}\text{H} \rightarrow \ ^{4}_{2}\text{He} + \ ^{1}_{0}\text{n}$$

   g. mass number: 40 = 40 + A, thus A = 0
   atomic number: 19 = 18 + Z, thus Z = 1
   the missing species is a positron.

$$^{40}_{19}\text{K} \rightarrow \ ^{40}_{18}\text{Ar} + \ ^{0}_{1}\beta$$

h. mass number: 55 + A = 55, thus A = 0
   atomic number: 26 + Z = 25, thus Z = -1
   the missing species is an orbital electron.

$$_{26}^{55}Fe \quad + \quad _{-1}^{0}\beta \quad \rightarrow \quad _{25}^{55}Mn$$

2. a. Molybdenum-96 has an even number of protons and neutrons. Technetium-96 has odd numbers of both protons and neutrons.

b. Nitrogen-14 is one of the few isotopes that is stable despite having odd numbers of both protons and neutrons. Phosphorus-30 is not one of those exceptions.

3. First it is necessary to determine the value of the rate constant for the radioactive decay of carbon-14.

$$k = \frac{0.693}{t_{1/2}} = \frac{0.693}{5730 \text{ yr}} = 1.209 \times 10^{-4} \text{ yr}^{-1}$$

Now substitute in the first order rate equation.

$$\ln \frac{N}{N_o} = - kt$$

$$\ln (0.673) = - (1.209 \times 10^{-4} \text{ yr}^{-1}) \times t$$

$$\underline{t = 3270 \text{ yr}}$$

This places the age of the ship within the range of dates given for the Late Bronze Age.

4. a. Each alpha particle represents the loss of four units of mass; each beta particle represents the loss of zero units of mass. Since five alpha particles were emitted, the total mass change is

$$\text{mass change} = 5 \times -4 = -20 \text{ amu}$$

Each alpha particle represents a decrease of two in the atomic number; each beta particle represents a gain of one unit in atomic number. Since five alpha particles and two electrons were emitted the total change in atomic number is

atomic number change = 5 x (-2) + 2 x 1 = -8

The new mass number is 238 - 20 = 218

The new atomic number is 92 - 8 = 84

Thus the product at this stage is $^{218}_{84}$Po

b. Since five alpha particles were emitted, the total mass change is

$$mass\ change = 5\ x\ -4 = -20\ amu$$

Since five alpha particles and two electrons were emitted the total change in atomic number is

atomic number change = 5 x (-2) + 2 x 1 = -8

The new mass number is 232 - 20 = 212

The new atomic number is 90 - 8 = 82

The product at this stage is $^{212}_{82}$Pb

5. Boron-10 consists of 5 protons and 5 neutrons, so the total predicted mass should be

$$5\ x\ 1.00783 + 5\ x\ 1.00867 = 10.08250$$

The mass defect is the difference between the actual and the predicted mass.

Mass defect = actual mass - predicted mass

Mass defect = 10.01294 g/mol - 10.08250 g/mol

Mass defect = - 0.06956 g/mol (or 6.956×10^{-5} kg)

Energy liberated = mass defect x c^2

$$= - 6.956 \times 10^{-5}\ kg/mol\ x\ (2.998 \times 10^8\ m/s)^2$$

Energy   = $- 6.26 \times 10^{12}$ J/mol

PRACTICE TEST (30 Min.)

1. Fill in the single missing species to complete each reaction.

a. $^{238}_{92}U \rightarrow \quad ^{4}_{2}He \; + \; \underline{\hspace{2cm}}$

b. $^{234}_{90}Th \rightarrow \underline{\hspace{2cm}} + \; ^{234}_{91}Pa$

c. $^{96}_{42}Mo \; + \; ^{2}_{1}H \rightarrow \underline{\hspace{2cm}} + \; ^{1}_{0}n$

d. $^{1}_{0}n \; + \; ^{235}_{92}U \rightarrow \; ^{143}_{55}Cs \; + \; 2 \; ^{1}_{0}n \; + \; \underline{\hspace{2cm}}$

e. $^{59}_{26}Fe \rightarrow \underline{\hspace{2cm}} + \; ^{59}_{27}Co$

f. $^{12}_{6}C \; + \; ^{12}_{6}C \rightarrow \; ^{1}_{1}H \; + \; \underline{\hspace{2cm}}$

g. $^{40}_{19}K \rightarrow \underline{\hspace{2cm}} + \; ^{0}_{+1}\beta$

2. Each of these pairs of nuclides consists of one stable isotope and one that is unstable. Select the isotope that is more likely to be stable in each case and explain your choice.

a. $^{11}_{6}C$ or $^{12}_{6}C$ b. $^{238}_{92}U$ or $^{197}_{79}Au$

3. Match the radioisotope on the left with the appropriate letter for its use from the right hand column.

_____ cobalt-60   a. used to treat and diagnose thyroid gland disorders

_____ iodine-125   b. a fissionable nuclei

_____ carbon-14   c. used to destroy cancer cells

_____ plutonium-239   d. used to "tag" compounds

126

4. Strontium-90, a product of the explosion of nuclear fission bombs, has a half-life of 28.8 years. (a) What fraction of the strontium-90 produced by an atmospheric nuclear test will still remain somewhere in the environment after a period of 50.0 years? (b) How long must pass before 90.0% of an initial amount of strontium-90 has vanished?

5. If a 10.0 grams sample of the radioactive isotope cesium-134 is allowed to remain undisturbed for 1.6 years, only 6.00 grams of the cesium is found to remain. What is the half-life of cesium-134?

CONCEPT TEST ANSWERS

1. helium nuclei, high energy electrons
2. gamma (γ)
3. sum of the mass numbers of the nuclei produced, the sum of the atomic numbers of the reactants.
4. bismuth

5. $_1^2H$, $_3^6Li$, $_5^{10}B$, and $_7^{14}N$ 6. $_{26}^{56}Fe$
7. activity   8. activity (or half-life)
9. binding energy   10. chain reaction
11. cadmium   12. 20%
13. hydrogen
14. neutron activation analysis
15, even, even

PRACTICE TEST ANSWERS

1. (L1)

a. $_{92}^{238}U \rightarrow _2^4He + _{90}^{234}Th$

b. $_{90}^{234}Th \rightarrow _{-1}^0\beta + _{91}^{234}Pa$

c. $_{42}^{96}Mo + _1^2H \rightarrow _{43}^{97}Tc + _0^1n$

d. $_0^1n + _{92}^{235}U \rightarrow _{55}^{143}Cs + 2 _0^1n + _{37}^{91}Rb$

e. $_{26}^{59}Fe \rightarrow _{-1}^0\beta + _{27}^{59}Co$

f. $^{12}_{6}C$ + $^{12}_{6}C$ → $^{1}_{1}H$ + $^{23}_{11}Na$

g. $^{40}_{19}K$ → $^{40}_{18}Ar$ + $^{0}_{+1}\beta$

2. **(L2)** a. Carbon-12 has an even number of both protons and neutrons, and so is favored over carbon-11, which has an even number of protons but an odd number of neutrons.

b. Gold-197 is more likely to be stable, since uranium-238 has an atomic number greater than that of bismuth (which is the element with the highest atomic number that does not undergo radioactive decay).

3. **(L5)** c. used to destroy cancer cells - cobalt-60

    a. used to treat and diagnose
       thyroid gland disorders   - iodine-125

    d. used to "tag" compounds   - carbon-14

    b. a fissionable nucleus   - plutonium-239

4. **(L3)** (a) The value of k is 2.41×10^{-2} yr^{-1}, and the fraction gone after 50.0 years would then be 0.300. (b) 95.7 years must pass before 90.0% of the strontium has decayed.

5. **(L3)** The rate constant is 0.32 yr^{-1} and the half-life is 2.2 years.

128

CHAPTER 8
ATOMIC STRUCTURE

LEARNING GOALS:

1. Much of our information about atomic structure
has been gained from observing the way in which
matter absorbs or emits electromagnetic radiation.
Therefore, to understand this chapter you must
learn the fundamental wave properties of
electromagnetic radiation, including the basic
equation

$$\lambda\upsilon = c$$

(Sec. 8.1)

2. In addition to the traditional wave picture of
electromagnetic radiation, scientists such as
Planck and Einstein proposed that electromagnetic
radiation had a particle nature, as defined by the
equation

$$E = h\upsilon$$

You should also understand the basic ideas of the
Quantum Theory, since it's also a crucial component
of the modern theory of atomic structure. (Secs.
8.2 & 8.3)

3. The study of atomic line spectra is a third
major component in the development of the theory of
atomic structure. You should be able to use the
Rydberg equation and understand how it served as a
basis for Bohr's picture of the atom.

$$\Delta E = -(Rhc/n_f^2) - (Rhc/n_i^2)$$

You should also be able to calculate the energies
of atomic energy states and spectral lines for
hydrogen gas. (Sec. 8.4)

4. The next step in the development of the quantum
theory was provided by de Broglie, who proposed
that there was a specific wavelength associated
with a moving electron. Understand how de
Broglie's ideas served as a basis for Heisenberg's

129

development of the uncertainty principle and ultimately served as the basis for quantum mechanics. You should also be able to use de Broglie's equation

$$\lambda = h/mv$$

(Sec. 8.5 & 8.6)

5. According to the ideas of quantum mechanics, quantum numbers play a key role in describing the energy of the electrons in an atom. It is essential that you thoroughly learn the rules that determine which combinations of quantum number values are permitted as well as the letter designations, s, p, d, f, etc. for the ℓ values. In the next chapter, you will see how these ideas provide a basis for the understanding of atomic structure. (Sec. 8.7)

IMPORTANT NEW TERMS:

amplitude (8.1)
Balmer series (8.3)
Bohr theory of the atom (8.4)
de Broglie wavelength (8.5)
electromagnetic radiation (8.1)
emission spectrum (8.4)
excited state (8.4)
frequency (8.1)
ground state (8.4)
hertz (8.1)
line spectrum (8.4)
Lyman Series (8.4)
nodal plane (8.7)
node (8.1)
orbital (8.7)
photoelectric effect (8.3)
photon (8.3)
Planck's constant (8.2)
probability density (8.7)
quantum (or wave) mechanics (8.6)
quantum numbers (8.7)
quantum (8.2)
Rydberg equation (8.43)
Schrödinger equation (8.7)
spectrometer (8.4)
standing or stationary wave (8.1)

stationary state (8.4)
subshell (8.7)
uncertainty principle (8.6)
wave function (8.7)
wavelength (8.1)

CONCEPT TEST

1. If you tie down a string at both ends and pluck it (like a guitar or violin), the type of wave motion or vibration that result is called a(n) _____.

2. The product of the wavelength multiplied times the _____ for any wave phenomena is equal to the velocity of propagation for the waves.

3. In a vacuum, the velocity of propagation of all electromagnetic radiation is _____.

4. Plank suggested that all energy gained or lost by an atom must be some integer multiple of a minimum amount of energy called a(n) _____.

5. Based on studies of the photoelectric effect, Einstein proposed that light could be described as having particle-like properties, due to the presence of particles called _____.

6. The lowest-energy stationary state of an atom is called its _____.

7. According to Bohr, atomic line spectra can occur only when electrons move between _____.

8. According to Heisenberg's uncertainty principle, if one attempts to simultaneously measure the position and momentum of an electron, the more exactly the position is measured, the greater will be the _____ in the momentum measurement.

131

9. In order to solve the Schrödinger equation for an electron in three dimensions, it is mathematically necessary to introduce three numbers having integer values called _____.

10. Each quantum mechanical wave function does not have a readily interpretable physical meaning, but the square of the wave function gives the _____ of finding the electron at a certain point.

11. Each quantum number has a slightly different meaning. **n, the principal quantum number,** is a measure of the _____ ; ℓ, **the angular momentum quantum number**, is related to the _____ of the electron orbitals, and **m_ℓ, the magnetic quantum number**, specifies in which _____ within a subshell the electron is located.

12. a. When **n** = 5, which of the following values are possible for ℓ, ℓ = -1, 0, 3, 5, or 6? _____

b. When ℓ = 3, which of the following values are possible for m_ℓ, m_ℓ = -2, 0, 3, or 4. _____

13. a. When ℓ = 3, the maximum number of orbitals of this type in a given electron shell is _____ (how many?), and a letter designation of _____ (choose from s,p,d, or f) is assigned.

b. When ℓ = 1, the maximum number of orbitals of this type in a given electron shell is _____ ,and a letter designation of _____ is assigned.

14. A region of a probability density graph where the probability of finding the electron is zero is called a(n) _____.

15. One nanometer is _____ meters.

STUDY HINTS:

1. As you may have noticed, a special feature of this chapter is the historical method of presentation. There is a reason for this. Often the experiments or equations developed by a given scientist are closely associated with his name, for example, the de Broglie equation or the Schrödinger equation. Therefore, it's important for you to remember the nature of the work done by each individual. This should also help you to visualize the complex set of developments in several different fields of science that all combined to produce the theory of atomic structure.

2. Students are sometimes confused by the reciprocal time units used for frequencies, but these units do make sense. By comparison, if someone asks you how often you eat regular meals, you may answer, "three times per day." If we write this statement in the same way as frequency results, it would become 3 day^{-1}. When we say that the frequency of a wave phenomena is 1,000,000 hertz, we are only saying that 1,000,000 waves per second would go past a stationary observation point.

3. Having read the chapter, you may well wonder whether an electron really is a wave or a particle. One way to satisfy yourself is to consider that we are not really arguing about the nature of an electron, but rather about which of two possible explanations works best. Our problem is apparently that neither explanation works all of the time, that is, the true nature of the electron is not exactly like either of our theories.

4. There is no substitute for carefully memorizing the rules that determine the possible quantum number values. It is impossible to understand atomic structure unless you have thoroughly learned these rules. In addition, it is essential to memorize the correspondence between the numerical values of ℓ and the letter values, i.e. s, p, d, f, etc. Until you know these relationships well, it will be difficult to follow any description of what

the quantum numbers mean.

PRACTICE PROBLEMS

1. (**L1**) a. For each pair of types of electromagnetic radiation given, circle the type having the higher energy.

ultraviolet or infrared

x-rays or γ-rays

blue light or yellow light

b. For each pair of types of electromagnetic radiation given circle the type having the greater wavelength.

microwave or infrared

AM radio or FM radio

x-rays or ultraviolet

2. (**L3**) According to the Bohr Theory, the seventh line of the Paschen series in the emission spectrum of hydrogen gas occurs when an electron goes <u>from</u> an orbit having an n value of _____ to an orbit having an n value of _____ .

3. (**L5**) a. When **n** = 5, what are the possible values for ℓ?

_____

b. When ℓ = 5, what are the possible values for m_ℓ?

_____

c. When **n** = 2, what are the possible values for m_ℓ?

_____

4. (**L5**) State whether each of the following combinations of quantum numbers is allowed or not according to the rules you learned in the text? If it is not allowed, explain why.

a. **n** = 3, ℓ = 2, and m_ℓ = -2 _____
b. **n** = 6, ℓ = 0, and m_ℓ = 1 _____
c. **n** = 2, ℓ = 1, and m_ℓ = 0 _____
d. **n** =-1, ℓ = 1, and m_ℓ = 1 _____

5. (**L5**) a. What is the maximum number of p orbitals that are found in a given electron shell of an atom? _____ b. What is the maximum number of d orbitals that are found in a given electron shell of an atom? _____

6. (**L4**) According to Heisenberg's uncertainty principle, we cannot simultaneously determine the position and momentum of an object. Why don't we observe this effect in our everyday life?

7. (**L1 and L2**) Calculate (a) the frequency and (b) the energy of a photon of light having a wavelength of 590 nanometers.

8. (**L4**) Calculate the de Broglie wavelength (in nanometers) of a golf ball moving with a speed of 145 km/s, if the mass of the golf ball is 45.7 grams.

9. (**L3**) Calculate the energy and the frequency of the fourth line in the Paschen Series of the emission spectrum of hydrogen atom in kJ/mole. The electrons that produce this series of lines move from higher energy states to the n = 3 state.

$(R = 1.0974 \times 10^7 \text{ m}^{-1})$

PRACTICE PROBLEM SOLUTIONS

1. a. The radiation having the higher energy is

   ultraviolet

   γ-rays

   blue light

   b. The radiation having the larger wavelength is

   microwave

   AM radio

   ultraviolet

2. from n = 10 to n = 3
3. a. +4, +3, +2, +1, and 0
   b. +5, +4, +3, +2, +1, 0, -1, -2, -3, -4, and -5
   c. $\ell = 1$, $m_\ell = +1$, 0, -1
      $\ell = 0$, $m_\ell = 0$
4. b. is not allowed
      because m_ℓ cannot be greater than ℓ
   d. is not allowed
      because n cannot have negative values
5. a. three p orbitals, b. five d orbitals
6. The uncertainty for objects that we encounter in everyday life is much smaller than our normal ability to detect. The uncertainty is still there; we just aren't aware of it.

7. a. Convert the wavelength to meters and substitute into the equation that relates wavelength and frequency.

$$\upsilon \;=\; \frac{c}{\lambda} \;=\; \frac{(3.00 \times 10^{8} \text{ m/s})}{(590 \text{ nm})(1 \times 10^{-9} \text{ m/nm})}$$

$$\upsilon \;=\; 5.08 \times 10^{14} \text{ s}^{-1}$$

b. Using this frequency in the Planck relationship

$$E = h\upsilon \;=\; (6.63 \times 10^{-34} \text{ J} \cdot \text{s})(5.08 \times 10^{14} \text{ s}^{-1})$$

$$E = 3.37 \times 10^{-19} \text{ J}$$

8. Convert the mass to kilograms and the speed to meters/second, then use the de Broglie wavelength equation

$$\text{wavelength of ball} = \lambda = \frac{h}{mv}$$

$$\lambda \;=\; \frac{(6.6262 \times 10^{-34} \text{ J} \cdot \text{s})(1000 \text{ g/kg})}{(45.7 \text{ g})(145 \text{ km/s})(1000 \text{ m/km})}$$

$$\lambda \;=\; 1.00 \times 10^{-37} \text{ m}$$

Convert to nanometers

$$\lambda \;=\; (1.00 \times 10^{-37} \text{ m})(1 \times 10^{9} \text{ nm/m})$$

$$\lambda \;=\; 1.00 \times 10^{-28} \text{ nm}$$

9. The energy difference is given by

$$\Delta E \;=\; -\,(Rhc/n_2^2) \;-\; (Rhc/n_1^2)$$

$$\Delta E \;=\; -\,(Rhc/7^2) \;-\; (Rhc/3^2)$$

$$\Delta E \;=\; -(1.0974 \times 10^{7} \text{ m}^{-1})(6.6261 \times 10^{-34} \text{ J} \cdot \text{s})$$
$$\times\; (2.9979 \times 10^{8} \text{ m/s})(-0.0907)$$

$$\Delta E \;=\; 1.9772 \times 10^{-19} \text{ J/atom}$$

137

Multiply by Avogadro's number and convert to kilojoules

$$\Delta E = (1.9772 \times 10^{-19} \text{ J/atom})$$
$$\times (6.022 \times 10^{23} \text{ atoms/mol})(1\text{kJ}/1000\text{J})$$

$$\underline{\Delta E = 119.07 \text{ kJ/mole}}$$

Now convert this energy into a frequency

$$\upsilon = \frac{\Delta E}{h} = \frac{(-119.07 \text{ kJ/mol})(10^3 \text{J/kJ})}{(6.022 \times 10^{23} \text{ electrons/mol})(6.626 \times 10^{-34} \text{ J} \cdot \text{s})}$$

$$\underline{\upsilon = 2.984 \times 10^{14} \text{ s}^{-1}}$$

PRACTICE TEST (40 min.)

1. Listed on the left below are the names of some of the scientists who made major contributions to the theory of atomic structure. On the right is a list of those contributions. Match the idea with the scientist by placing the letter from the right hand column in the appropriate space on the left.

_____ Balmer

_____ Bohr

_____ de Broglie

_____ Einstein

_____ Heisenberg

_____ Thomson

_____ Planck

a. position and momentum cannot be measured exactly and simultaneously

b. The wave properties of the matter

c. explanation of atomic spectra

d. mass to charge ratio for the electron

e. explained the photoelectric effect

f. quantization of energy emission and absorption

g. mathematical relation that predicts the first three lines of the visible emission spectrum for hydrogen gas

138

2. According to the Bohr Theory, the fifth line of the Lyman series in the emission spectrum of hydrogen gas occurs when an electron goes _from_ an n value of _____ to an n value of _____.

3. Match the terms in the left hand column with the most appropriate description from the right hand column.

_____ photoelectric effect   a. only certain quantized atomic energy levels are allowed

_____ nodal plane   b. electrons are ejected when light strikes the surface of some metals

_____ stationary states   c. the more exactly the momentum of a particle is known, the less exactly the position can be measured

_____ wave function   d. region where the probability of finding an electron is zero

_____ uncertainty principle   e. solutions of the Schrödinger equation

4. a. What is the maximum number of p orbitals that are found in a given electron shell of an atom? b. What is the maximum number of g orbitals?

5. Which of the following combinations of quantum numbers is not allowed according to the rules discussed in the textbook.

a. $n = 0$, $\ell = 2$, and $m_\ell = 1$.
b. $n = 4$, $\ell = 0$, and $m_\ell = 0$.
c. $n = 3$, $\ell = 2$, and $m_\ell = 3$.
d. $n = 5$, $\ell = -4$, and $m_\ell = 0$.

6. If the wavelength is 7.00×10^{-5} cm for a certain photon of light, what is the energy of this photon in joules?

7. Infrared radiation has a wavelength of 1.2×10^3 nanometers.   a. What is the frequency of this radiation? b. What is the energy in joules of one photon of this radiation? c. What is the energy in joules of 1.00 moles of photons of this radiation?

8. If an electron has a mass of 9.109×10^{-31} kilograms and is moving with a velocity 10.0% of the velocity of light, what is its de Broglie wavelength?

9. The Pfund Series of atomic spectral lines for hydrogen gas are the result of electronic transitions that terminate in the energy level where n = 5. What is the energy of the fourth line of the Pfund Series in J/atom? in kJ/mole?

CONCEPT TEST ANSWERS

1. standing or stationary wave
2. frequency   3. 2.998×10^8 m/s
4. quantum   5. photons
6. ground state   7. stationary states
8. uncertainty   9. quantum numbers
10. probability (or intensity of the electron wave)
11. orbital size or diameter, shape, orbital
12. a. 0 and 3 for ℓ,
    b. -2, 0, and 3 for m_ℓ.
13. a. seven f orbitals
    b. three p orbitals
14. node   15. 1×10^{-9} meters

PRACTICE TEST ANSWERS

1. Balmer - g, Bohr - c, de Broglie - b, Einstein - e, Heisenberg - a, Thomson - d, and Planck - f.
2. (L3) from n = 6 to n = 1
3. (L2, L4, and L5)

photoelectric effect -   b. electrons are ejected
                            when light strikes a metal
                            surface

nodal plane -   d. region where the
                            probability of finding an
                            electron is zero

stationary states -   a. only certain quantized
                            atomic energy levels are
                            allowed

wave function -   e. solutions of the
                            Schrödinger equation
uncertainty principle -   c. the more exactly the
                            momentum of a particle is
                            known, the less exactly
                            the position can be
                            measured

4. **(L6)** a. $\ell = 1$, so three orbitals are possible.
   b. $\ell = 4$, so nine orbitals are possible.
5. **(L6)** The combinations not allowed are
   a. (**n** cannot equal zero),
   c. (**m_ℓ** cannot be greater than ℓ.), and
   d. (ℓ cannot have a negative value.)
6. **(L2)** 2.84×10^{-19} J
7. **(L1 and L2)** a. 2.5×10^{14} s^{-1}

   b. 1.7×10^{-19} joules   c. 1.0×10^{5} joules
8. **(L4)** 2.426×10^{-11} meters
9. **(L3)** 6.028×10^{-20} J/atom, 36.30 kJ/mol

CHAPTER 9
ATOMIC ELECTRON CONFIGURATIONS
AND PERIODICITY

LEARNING GOALS:

1. In addition to reviewing the three quantum numbers discussed in the previous chapter, you should also understand the use of the fourth quantum number, m_s, which is related to the spin magnetic moment of the electron. These quantum numbers and the Pauli exclusion principle are fundamental to the theory of atomic electronic configuration. (Sec. 9.1 & 9.2)

2. In order to understand atomic structure, it is essential to learn the various ways of remembering the relative energy order of the atomic orbitals. The most obvious result of these relative energies is the shape of the periodic table. (Sec. 9.3)

3. You should be able to depict the electronic configuration of any element or simple monatomic ion by
   (a) writing a set of four quantum number values for each electron in the atom,
   (b) using orbital box diagrams, or
   (c) using spectroscopic notation.
In each of these cases, the notation can be simplified by using the appropriate rare gas configuration. (Sec. 9.4 & 9.5)

4. Be able to use the periodic table to organize the periodic trends in properties such as atomic radius, ionization energy, electron affinity, and ionic size when comparing elements across a period or down a group. (Sec. 9.6)

IMPORTANT NEW TERMS:

actinide elements (9.4)
atomic size (9.6)
core electrons (9.3)
diamagnetic (9.1)

electron affinity (9.6)
electron spin magnetic quantum number (9.1)
ferromagnetic (9.1)
Hund's rule (9.4)
ion size (9.6)
ionization energy (9.6)
isoelectronic (9.6)
lanthanide elements (9.4)
law of chemical periodicity (9.6)
orbital box notation (9.2)
paramagnetic (9.1)
Pauli exclusion principle (9.2)
pseudo-rare gas configuration (9.7)
rare gas notation (9.4)
spectroscopic notation (9.4)
valence electrons (9.3)

CONCEPT TEST

1. Substances that are weakly attracted to a magnetic field but loose their magnetism when removed from the magnetic field are called

_____.

2. Each atomic orbital can be assigned (or contain) a maximum of how many electrons? _____

3. What is the maximum number of electrons possible in an electronic shell having an $n = 4$? _____

4. For each pair, circle the orbital that lies lower in energy. (Hint: Compare the $n + \ell$ value)

(a) 4s or 3d   (b) 4p or 5s   (c) 4f or 5d?

5. Substances that maintain their magnetism when withdrawn from a magnetic field are called

_____.

6. According to Hund's rule, when electrons are assigned to different orbitals in the same subshell, the most stable arrangement is that with the maximum number of _____.

143

7. The energy needed to remove an electron from an atom in the gas phase is called _____.

8. The electron configurations of stable monatomic positive or negative ions are described as

_____ or _____ configurations.

9. Which of the following best describes the variation of atomic radii of the elements with respect to their position on the periodic table?

a. decreases across a period, increases down a group.
b. decreases across a period, decreases down a group.
c. increases across a period, increases down a group.
d. increases across a period, decreases down a group.

10. Which type of element below has atomic sizes that remain almost identical across a period.

a. main group metals   b. main group nonmetals
        c. transition metal elements

11. Circle any elements in the following list that are p-block elements.

        Cu   Na   Cl   Cr   Al

12. Circle any elements in the following list that are d-block elements.

        Cu   Na   Cl   Cr   Al

13. For each of the following elements, indicate which rare gas symbol would be used if we wished to abbreviate the electronic configuration.

Ag _____   Ti _____

Ba _____   Fe _____

14. Which of the following best describes the variation of ionization energy of the elements with respect to their position on the periodic table?

a. decreases across a period, increases down a group.
b. decreases across a period, decreases down a group.
c. increases across a period, increases down a group.
d. increases across a period, decreases down a group.

15. Electron spin is quantized is such a way that, in an external magnetic field, _____ (give number) orientations of the electron spin are possible.

STUDY HINTS:

1. The best way to remember the filling order of the atomic orbitals is to understand the relationship between the periodic table and the atomic electronic configurations. As you write electronic configurations, compare them with the periodic table and try to see how the table is simply the result of the filling of the different types of orbitals.

Another way to remember the filling order is to construct the diagram on the right. Simply list the possible orbitals in the order shown, and then draw diagonal lines from upper right to lower left to show the filling order.

```
 1s
 2s  2p
 3s  3p  3d
 4s  4p  4d  4f
 5s  5p  5d  5f  5g
 6s  6p  6d  6f  6g
```

2. Normally we use a filling order chart, like that in the previous hint, to predict what the periodic table will look like. Actually the best assistance to remembering the electronic configurations is the table itself. With a little practice, you will find that the table itself is the best possible aide to remembering most of the orbital filling orders.

3. Remember that both the periodic chart and the diagram above only give the filling order of the orbitals. For instance, the 4s fills before the 3d, but once both orbitals are filled the 3d is lower in energy than the 4s. That is, the electron configuration of germanium would be written as

$$1s^2 2s^2 2p^6 3s^2 3p^6 3d^{10} 4s^2 4p^2$$

rather than $1s^2 2s^2 2p^6 3s^2 3p^6 4s^2 3d^{10} 4p^2$.

Check to determine which way your instructor wishes you to write the spectroscopic notation.

4. Remember that the law of chemical periodicity is based on atomic numbers, not atomic masses. Throughout most of the periodic chart, the elements seem to be arranged in order of increasing atomic mass, but this is not always true. Can you find three places on the periodic table where the order of the elements is the reverse of that expected from the atomic masses?

5. Be sure that you recognize the difference between electron affinity and ionization energy, especially the sign on the energy change! It always requires energy to ionize an atom, but energy is always released when an electron is added to form a stable anion.

PRACTICE PROBLEMS

1. (**L3**) Circle the following ions which are <u>not</u> likely to be found in chemical reactions under normal conditions?

Li^+ S^{3-} Sc^{4+} Mg^{2+} Ne^+ Br^-

2. (**L3**) a. Which element is the first p block element? _____
b. Which element is the first d block element? _____

3. (**L3**) Write electron configurations for each of these elements on the next page using both orbital box diagrams and spectroscopic notation. Also indicate for each element whether it is paramagnetic or diamagnetic. You may use rare gas notation to simplify your answer.

(a) magnesium (atomic number = 12)

(b) phosphorus (atomic number = 15)

(c) cobalt (atomic number = 27).

4. (**L4**) Arrange the following elements in order of increasing first ionization energy: Cs, C, K, Li, and F.

5. (**L4**) Arrange the following ions in order of increasing size: O^{2-}, N^{3-}, and F^-.

6. (**L3**) Use spectroscopic notation to represent the electron configuration of the following elements. Indicate for each element whether or not it is paramagnetic. Use rare gas notation if you wish.

(a) The element zirconium, atomic number = 40, is found in a number of minerals that have been known since biblical times, but the pure element wasn't isolated until 1914.

(b) The radioactive element polonium, atomic number = 84, was the first element discovered by Mme. Curie and was named for Poland, her native country.

(c) Iodine, element 53, can cause health problems either by being present or being absent. A small amount of iodine in the diet is necessary to prevent goiter, but iodine vapor is irritating to the eyes and can cause lesions on the skin.

7. (**L3**) Write down a complete set of four quantum numbers for each of the electrons beyond the nearest rare gas for the elements
(a) cerium (atomic number = 58)

(b) scandium (atomic number = 21).

8. (**L1**) What is the maximum number of electrons that can be identified with each of the following sets of quantum numbers? If the answer is none, explain why this is so.

(a) $n = 2$, $\ell = 2$, $m_\ell = 0$ _____

(b) $n = 5$, $\ell = 2$ _____

(c) $n = 6$, $\ell = 2$, $m_\ell = 1$, $m_s = +1/2$ _____

9. (**L4**) The electron configurations of A and B, two unknown elements, are

$$A = \ldots \; 3s^2 3p^5 \qquad\qquad B = \ldots \; 3s^2 3p^2$$

a. Indicate whether each element is a metal, a metalloid, or a nonmetal.

A _____ B _____

b. Which of these two elements is expected to have a smaller atomic radius? _____

c. Predict the formula of a likely compound formed only by these two elements. _____

10. (**L4**) Name each of the following elements based on the information provided.

a. electronic configuration is
$$1s^2 2s^2 2p^6 3s^2 3p^6 4s^2 3d^8$$ _____

b. the alkali metal with the largest ionization energy _____

c. the element whose +4 ion has the configuration $[Kr]4d^3$ _____

d. the first element (that is the one with the lowest atomic number) that has f electrons in its ground state electronic configuration. _____

11. (**L3**) Use spectroscopic notation (and the appropriate noble gas symbol) to represent the electron configuration for each of the following ions:

a. Ni^{2+}

b. S^{2-}

c. Mo^{3+}

d. Pt^{2+}

PRACTICE PROBLEM SOLUTIONS

1. The following ions do not have a noble gas configuration, and so are not expected to be found in nature under normal conditions: S^{3-}, Sc^{4+}, and Ne^+.

2. Boron is the first element (i.e. the one with the smallest atomic number) that has p electrons,

and scandium is the first d block element.
3. a. magnesium - diamagnetic

(↑↓) (↑↓)　(↑↓) (↑↓) (↑↓)　(↑↓)
 1s　2s　　　　　2p　　　　3s

spectroscopic notation: $1s^2 2s^2 2p^6 3s^2$

a. phosphorus - paramagnetic

(↑↓)　(↑↓)　(↑↓) (↑↓) (↑↓)　(↑↓)　(↑)(↑)(↑)
 1s　　2s　　　　2p　　　　3s　　　　3p

spectroscopic notation: $1s^2 2s^2 2p^6 3s^2 3p^3$

b. cobalt - paramagnetic
 [Ar] (↑↓)　　　　(↑↓) (↑↓) (↑)(↑)(↑)
　　　4s　　　　　　　　3d

spectroscopic notation: $1s^2 2s^2 2p^6 3s^2 3p^6 4s^2 3d^7$

4. Cs < K < Li < C < F

5. $F^- < O^{2-} < N^{3-}$

6. a. zirconium [Kr] $5s^2 4d^2$ paramagnetic
or
$1s^2 2s^2 2p^6 3s^2 3p^6 3d^{10} 4s^2 4p^6 5s^2 4d^2$

b. polonium [Xe] $6s^2 4f^{14} 5d^{10} 6p^4$ paramagnetic
or
$1s^2 2s^2 2p^6 3s^2 3p^6 3d^{10} 4s^2 4p^6 4d^{10} 4f^{14} 5s^2 5p^6 5d^{10} 6s^2 6p^4$

c. iodine [Kr] $5s^2 4d^{10} 5p^5$ paramagnetic
or
$1s^2 2s^2 2p^6 3s^2 3p^6 3d^{10} 4s^2 4p^6 4d^{10} 5s^2 5p^5$

7. a.

n	ℓ	m_ℓ	m_s
6	0	0	+1/2
6	0	0	−1/2
5	+2	+2	+1/2
4	+3	+3	+1/2

Other combinations are possible for the d and f electrons.

b.	n	ℓ	m_ℓ	m_s
	4	0	0	+1/2
	4	0	0	-1/2
	3	+2	+2	+1/2

Other combinations are possible for the 3d electron.

8. a. none, **n** and ℓ cannot have the same value.
   b. This corresponds to a set of five 5d orbitals, and so may contain a maximum of 10 electrons.
   c. This corresponds to a single 6d electron.

9. a. A is a nonmetal; B is a metalloid.
   b. A should have the smaller atomic radius.
   c. BA_4

10. a. nickel
    b. lithium
    c. technetium
    d. cerium

11. a. Ni^{2+} $[Ar]3d^8$
    b. S^{2-} $[Ne]3s^2 3p^4$
    c. Mo^{3+} $[Kr]4d^3$
    d. Pt^{2+} $[Xe]4f^{14}5d^8$

PRACTICE TEST (40 Minutes)

1. Give the symbols of all of the elements that in their ground states have (a) three p electrons in their outermost subshell and (b) two d electrons in their outermost subshell.

2. List all of the elements in the first three periods of the periodic table that have two unpaired electrons in their ground state.

3. Write electron configurations for the following elements using both orbital box diagrams and spectroscopic notation: (a) K and (b) As. Indicate for each element whether or not it is paramagnetic.

4. Arrange the following elements in order of decreasing atomic radius: Si, Cl, Mg, Sr, and F.

5. Arrange the following elements and ions in order of decreasing size: Se^{2-}, Kr, Br^-, and As^{3-}.

6. What is the maximum number of electrons that can be identified with each of the following sets of quantum numbers? If the answer is none, explain why this is so.

(a) $n = 3$

(b) $n = 3$, $\ell = 2$, $m_\ell = -2$

(c) $n = 5$, $\ell = 3$, $m_\ell = -3$, $m_s = 0$

7. Name each of the following elements based on the information provided.

a. the electronic configuration is $[Ar]4s^23d^8$
b. the element whose +3 ion has the configuration $[Ar]3d^2$
c. the element in Group 5A with the highest first ionization energy.
d. the element in the third period with the greatest number of d electrons.

8. Use spectroscopic notation to represent the electron configuration of the following elements. Indicate for each element whether or not it is paramagnetic. You may use the rare gas notation if you wish.

a. The element palladium, atomic number = 46, is used for dentistry, watch and instrument making, and also as a catalyst.

b. Gallium, atomic number = 31, is one of the few metals that is a liquid at about room temperature. (Can you name another metal that is a liquid?)

c. Rhenium, atomic number = 75, is a very dense element with a high melting point used as a catalyst and also in many different alloys, where it increases corrosion and wear resistance.

9. Use Spectroscopic notation to represent the electron configuration for each of the following ions:

a. Zn^{2+} b. P^{3-} c. Nb^{3+}

10. Write down a complete set of four quantum numbers for each of the electrons beyond the nearest rare gas for the elements:
(a) manganese (atomic number = 25) and (b) cadmium (atomic number = 48)

11. Suppose that a friend of yours is writing a science fiction story involving an alternative universe where the quantum number rules are slightly different from those that are actually known to be true. He proposes to make one small change; the values of m_s can be +1/2, -1/2, and 0. **All other quantum number rules will remain exactly the same.** To help him understand the results of this change and make sure the story is consistent (even though it is only an imaginary universe), answer the following questions.
a. What is the maximum number of electrons allowed in a single orbital?
b. How many elements are possible in the first period, that is, having **n** =1?
c. How many elements are possible in the second period, that is, having **n** = 2?
d. How many elements are possible in the third period, that is, having **n** = 3?
e. What is the electron configuration of an element having an atomic number of 5? Will this element be a metal or a nonmetal? What is a probable oxidation state for this element?
f. What is the electron configuration of an element having an atomic number of 26? Will this element be a metal or a nonmetal? What is a probable oxidation state for this element?
g. If the element described in part e has the symbol X, and the element described in part f has the symbol Z, give the formula for a compound formed by the combination of these two elements.

CONCEPT TEST ANSWERS

1. paramagnetic
2. two
3. $2 \times n \times n = 32$
4. 4s, 4p, & 4f are lowest.
5. ferromagnetic
6. unpaired electrons
7. ionization energy
8. rare gas or pseudo-rare gas
9. a. decreases across a period, increases down a group.
10 c. transition metal elements
11. Al and Cl
12. Cu and Cr
13. Ag <u>Kr</u>   Ti <u>Ar</u>   Ba <u>Xe</u>   Fe <u>Ar</u>

14. d. increases across a period, decreases down a group.
15. two

PRACTICE TEST ANSWERS

1. (L3) a. N, P, As, Sb, and Bi
   b. Ti, Zr, and Hf

2. (L3) carbon, silicon, oxygen, and silicon

3. (L3) a. potassium - paramagnetic

   [Ar] ($\uparrow$)
        4s

spectroscopic notation: $1s^2 2s^2 2p^6 3s^2 3p^6 4s^1$

b. arsenic - paramagnetic

[Ar] ($\uparrow\downarrow$)   ($\uparrow\downarrow$)($\uparrow\downarrow$)($\uparrow\downarrow$)($\uparrow\downarrow$)($\uparrow\downarrow$)   ($\uparrow$)($\uparrow$)($\uparrow$)
    4s   3d   4p

spectroscopic notation: $1s^2 2s^2 2p^6 3s^2 3p^6 4s^2 3d^{10} 4p^3$

4. (L4) Sr > Mg > Si > Cl > F

5. (L4) $As^{3-} > Se^{2-} > Br^- > Kr$

6. **(L1)** a. The n = 3 set of orbitals may contain 18 electrons.
b. A single 3d orbital may contain two electrons.
c. Because $m_s = 0$, this is not possible.

7. **(L3)** a. nickel   b. vanadium
c. nitrogen   d. zinc

8. **(L3)** a. $[Kr]5s^2 4d^8$ – paramagnetic
b. $[Ar]4s^2 3d^{10} 4p^1$ – paramagnetic
c. $[Xe]6s^2 4f^{14} 5d^5$ – paramagnetic

9. **(L3)** a. $[Ar]3d^{10}$ b. $[Ne]3s^2 3p^6$ c. $[Kr]4d^2$

10. **(L3)** a. manganese

n	ℓ	m_ℓ	m_s
4	0	0	+1/2
4	0	0	-1/2
3	+2	+2	+1/2
3	+2	+1	+1/2
3	+2	0	+1/2
3	+2	-1	+1/2
3	+2	-2	+1/2

a. cadmium

n	ℓ	m_ℓ	m_s
5	0	0	+1/2
5	0	0	-1/2
4	+2	+2	+1/2
4	+2	+1	+1/2
4	+2	0	+1/2
4	+2	-1	+1/2
4	+2	-2	+1/2
4	+2	-2	-1/2
4	+2	-1	-1/2
4	+2	0	-1/2
4	+2	+1	-1/2
4	+2	+2	-1/2

11. **(L3)** a. three   b. three

c. twelve   d. twenty seven

e. $1s^3 2s^2$, metal, +2

f. $1s^3 2s^3 2p^9 3s^3 3p^5$, nonmetal, −1

g. XZ_2

CHAPTER 10
BASIC CONCEPTS OF CHEMICAL BONDING
AND MOLECULAR STRUCTURE

LEARNING GOALS:

1.   Chemical bonding involves the atomic valence electrons, and so in order to understand bonding it's necessary to know how to identify the valence electrons of each element. The differences between covalent and ionic bonding, lone pair and bond-pair orbitals are also important ideas to master at this stage. (Secs. 10.1 & 10.2)

2. Although true ionic bonding is uncommon, be sure to know the properties of the elements involved that favor more ionic bonding, such as the electronegativity values or the position on the periodic table. Also be aware of the effect that greater ionic character has on properties such as melting point and solubility. (Secs. 10.3 & 10.4)

3. Lewis electron dot structures are one of the oldest methods of describing covalent bonding, but they are still widely used. Learn to draw Lewis dot structures for simple molecules and ions, in addition to being able to categorize bonds as single, double, or triple, and sigma or pi.   When using the Lewis approach, know how to represent cases where the octet rule is not followed, including odd-electron compounds, expanded valence compounds, and compounds where the central atom has less than eight valence electrons. (Sec. 10.4)

4. Molecules that have resonance structures or those that have a central atom with expanded valence are simply special cases of Lewis electron dot representation. Know how to represent these species. (Sec. 10.4)

5. Understand the major terms used to describe a chemical bond, including bond order, bond length, and bond energy. When appropriate data is provided, be aable to estimate bond energies from

157

enthalpies of formation, or vice versa. (Sec. 10.5)

6. Study carefully the rules in Section 10.5 that show how to determine the polarity of a chemical bond and understand the important role that electronegativity plays in polarity. Understand how to determine the atom formal charge for the atoms in either a molecule or an ion. (Sec. 10.5)

7. The three-dimensional structure of molecules provides important information that may be used for several different purposes, including the determination of chemical reactivity. The Valence Shell Electron Pair Repulsion (VSEPR) theory provides a useful representation of molecular structure. Learn the structural-pair geometry associated with two to six pairs of bonding electrons and use this information to predict the molecular geometry for molecules that have lone pairs as well as those that do not. Know how to use these predicted structures to estimate the approximate bond angles in these molecules. (Sec. 10.6)

8. When electronegativity differences causes a molecule to have a positive and a negative pole, the molecule is said to be a dipole. Understand how to recognize a dipolar molecule by combining information about molecular structure with the concept of electronegativity difference. (Sec. 10.7)

IMPORTANT NEW TERMS:

atom formal charge (10.5)
axial position (10.6)
bond orbital (10.2)
bond length (10.5)
bond polarity (10.5)
bond dissociation energy (10.5)
bond angle (10.6)
bond order (10.5)
chemical bond (10.2)
coordinate covalent bond (10.4)
covalent bond (10.2)
dipole moment (10.7)

double bond (10.4)
electronegativity (10.5)
equatorial position (10.6)
expanded valence (10.4)
ionic bond ((10.2)
Lewis electron dot symbol (10.1)
lone pair orbital (10.2)
molecular geometry (10.6)
nonpolar covalent bond (10.5)
octet rule (10.1)
odd-electron molecules (10.4)
pi (π) bond (10.4)
principle of electroneutrality (10.5)
resonance (10.4)
sigma (σ) bond (10.4)
single bond (10.4)
structural pairs (10.6)
structural-pair geometry (10.6)
triple bond (10.4)
valence electrons (10.1)
Valence Shell Electron Pair Repulsion Theory (10.6)

CONCEPT TEST

1. The number of valence electrons of each element is equal to that element's _____ on the periodic table.

2. When a chemical bond involves the complete transfer of one or more electrons from one atom to another, the resulting bond is said to be _____; when the electrons are shared the bond is said to be _____.

3. The oxidation number of an atom is the charge that the atom would have if ALL of its bonds were considered to be completely _____.

4. When writing Lewis structures, the central atom is the atom having the lowest _____ .

5. In Lewis structures, hydrogen is almost always attached to _____ other atom. Multiple bonds are most often formed by atoms of the following elements: _____, _____, _____, and _____ .

6. Complete the expression

Atom formal charge = _____

                  - the number of lone pair electrons

                  - ____ (number of bonding electrons)

7. The number of bonding electron pairs shared by two atoms in a molecule is called the _____ .

8. There is a triple bond between carbon and oxygen in a carbon monoxide molecule. This means that the bond consists of _____ sigma bonds and _____ pi bonds, and that the overall bond order is _____ .

9. List these pairs of bonded atoms in order of <u>increasing</u> bond length: O-F, O-I, and O-Cl.

10. The process of breaking bonds in a molecule is _____. (Circle the best answer.)

a. always exothermic   b. always endothermic

c. may be exothermic or endothermic depending on conditions.

11. Arrange the following pairs of bonded atoms in order of _increasing_ bond polarity: C-Cl, Na-Cl, and C-C. For polar bonds, underline the more positive atom.

12. The principle of electroneutrality states that for each atom in a molecule (or ion) the formal charge should be as close to a value of _____ as possible.

13. List the expected structural-pair geometry for each number of structural electron pairs.

(a) 6 _____

(b) 3 _____

(c) 5 _____

(d) 2 _____

14. The VSEPR model does not apply to those molecules where the central atom is a _____ _____.

15. The experimental measure of the separation of charge in a polar molecule is called the _____ _____.

STUDY HINTS:

1. Students sometimes make the job of determining the number of sigma and pi bonds much too complicated. You can count on the fact that the first pair of electrons bonding two atoms together

will always be sigma. Each additional pair of
bonding electrons will be a pi bond. For example,
consider the formaldehyde molecule

$$\begin{array}{c} O \\ \parallel \\ H - C - H \end{array}$$

It has one sigma bond between each carbon and
hydrogen. The carbon-oxygen double bond consists
of two electron pairs, one of which must be sigma
and the other pi.

2. When no single Lewis structure will adequately
represent a molecule, it is sometimes necessary to
use a combination of several structures called
resonance structures. The actual structure may be
thought of as a combination of all of these
resonance forms. Remember that these are
alternative pictures of the same molecule, so they
only differ in the distribution of electron pairs.
All resonance structures of the same molecule must
have the same atoms in the same positions, bonded
to the same partners.

3. Many students initially have difficulty
visualizing three-dimensional structures. If you
have this problem, it may be helpful to work with a
set of three-dimensional models. Practice by
trying to visualize the molecules shown in your
textbook. In each case first look at the formula
on the page and try to mentally picture how the
model will look. Then use the model to see how
closely your mental image corresponds to the real
thing. Some individuals will find this easier than
others, but practice will help anyone to improve
this skill.

4. It is important for you to distinguish between
the structural-pair geometry and the molecular
geometry. If a molecule has no electron lone
pairs, the arrangement of the atoms in space must
be identical with the arrangement of the electron
pairs. For molecules that do have lone pairs, the
structural-pair geometry will always be different
from the molecular geometry. Even in this case,
however, the two are related, since the position of
the atoms must be determined by the position of the

162

bonding electron pairs. The best approach is to first determine the structural-pair geometry, then determine how many lone pairs are present, and finally predict the molecular geometry based only on the arrangement of the bonding electron pairs.

5. It is important to remember that the lone pairs in a trigonal bipyramidal structural-pair geometry will be in the equatorial positions, otherwise you will have difficulty predicting the correct molecular geometry for these cases. If you are not sure of the difference between the axial and equatorial positions, use your molecular models to clarify this point.

6. There is one important limitation that you should remember when using the VSEPR Theory. Although it does an excellent job of predicting the structure of compounds that have a typical metal or a typical nonmetal as the central atom, it is frequently incorrect for compounds that have a transition metal as the central atom.

7. Remember that even though some or all of the chemical bonds in a molecule may be polar, the molecule can still be nonpolar because of the molecular geometry.

PRACTICE PROBLEMS

1. (L1) Give the periodic group number and the number of valence electrons for each of the following atoms:
    (a) Si   (b) F   (c) Al   (d) Li   (e) Be

group
number   _____   _____   _____   _____   _____

no. of
valence
electrons   _____   _____   _____   _____   _____

2. (L1) Which of the following atoms could have an expanded valence shell and form compounds with five or more valence electron pairs?

(a) S   (b) Si   (c) N   (d) O   (e) Br

163

3. (**L3**) Draw Lewis dot structures for the following molecules or ions:

(a) Cl_2 (b) CO_2

(c) NCl_3 (d) GeO

(e) ClO_2^- (f) NO_2^+

4. (**L1**) What is the bond order of each chemical bond in the following molecules. In each case, indicate the number of sigma and the number of pi bonds present.

```
              O                              H   O
              ‖                              |   ‖
(a)  H - C - O - H             (b)  H - C - C - H
                                             |
                                             H
```

5. (**L4**) The following molecules or ions have two or more resonance structures. Show all of the resonance structures for each species.

(a) SeO_3

164

(b) NO_3^-

6. (**L5**) Arrange the following pairs of bonded atoms in terms of increasing bond length:
(a) N-Cl, N-Br, and N-F

_____ < _____ < _____

(b) N-P, N-N, and N-As

_____ < _____ < _____

7. (**L5**) The combination of hydrogen and oxygen gases, a very energetic process that can be used to power a rocket, is described by the equation

$$2 \; H_2(g) \quad + \quad O_2(g) \quad \rightarrow \quad 2 \; H_2O(g)$$

Using the bond energies below, estimate the standard heat of combustion per mole of hydrogen gas.

Bond	Bond Energy
H – H	436 kJ/mol
O = O (O_2)	498
H – O	464

8. (**L2**) Arrange the following pairs of bonded elements in order of increasing ionic character of the C – X bond:

(a) C – N   (b) C – O   (c) C – I   (d) C – F

_____ < _____ < _____ < _____

165

9. (**L7**) The column on the left represents general formulas for several different types of simple molecules or ions. The letter M represents the central atom; B represents bonding electron pairs, and E represents lone pairs of electrons. For example, NF_3 would be represented MB_3E, since there are three bonding pairs and one lone pair on the nitrogen. The right hand column lists phrases that describe some of the possible molecular shapes that can exist. Match the general formula on the left with the appropriate molecular shape on the right by placing the letter from the right hand column in the space provided.

_____ MB_2E		a. square planar
_____ MB_5E		b. bent ($120°$)
_____ MB_3E		c. "T-shaped"
_____ MB_3E_2		d. trigonal pyramidal
_____ MB_2E_3		e. square pyramidal
_____ MB_4E_2		f. linear

10. (**L7**) Describe the central atom structural-pair shape and the molecular (or ionic) shape (including the <u>approximate</u> bond angle for bent molecules) for each of the following:

formula	structural-pair shape	molecular or ionic shape
a. BeF_2		
b. CCl_4		
c. BCl_3		
d. OF_2		
e. $BrCl_3$		
f. BCl_2^-		

11. (**L2**) Arrange the following compounds, CsBr, CsCl, and CsI, in order of increasing melting temperature.

Arrange the same series of compounds in order of increasing solubility in water.

12. (**L6**) Draw the Lewis structures and determine the atom formal charges for each atom in

(a)   hydrogen cyanide, HCN

(b) phosphorus oxychloride, $POCl_3$

PRACTICE PROBLEM SOLUTIONS

1.

element	group number	number of valence electrons
Si	4	4
F	7	7
Al	3	3
Li	1	1
Be	2	2

2. Sulfur, silicon, and bromine are each in the third or fourth period and so may have compounds with an expanded octet of electrons. Nitrogen and oxygen are second period elements and so cannot have an expanded octet.

3. a. (**1**) Since only two atoms are present, there isn't really a central atom.
(**2**) The total number of valence electrons is the sum of the group numbers, that is, since the group

167

number for chlorine is seven, two times 7 = 14. There are seven valence pairs.

(3) Place a single bond between each pair of atoms.

$$Cl - Cl$$

(4) and **(5)** The remaining six pairs are distributed around the two chlorines as lone pairs.

$$: \overset{..}{\underset{..}{Cl}} - \overset{..}{\underset{..}{Cl}} :$$

Each chlorine now has an octet, so the structure is complete. Notice that there is a single, sigma bond between the chlorines.

3. b. **(1)** Carbon has a lower electron affinity than oxygen, so carbon is the central atom.
(2) The total number of valence electrons is four from carbon and two times six from oxygen, or 16 electrons.
(3) Place a single bond between each pair of atoms.

$$O - C - O$$

(4) and **(5)** The remaining 12 electrons are distributed around the terminal oxygens, until each has four electron pairs.

$$: \overset{..}{\underset{..}{O}} - C - \overset{..}{\underset{..}{O}} :$$

(6) The carbon has a deficiency of two electron pairs, and so you must convert one lone pair from each of the oxygens into a pi bond between that oxygen and the carbon.

$$: \overset{..}{O} = C = \overset{..}{O} :$$

Each atom now has a share in four electron pairs. Notice that there are two sigma and two pi bonds.

3. c. **(1)** Nitrogen is the central atom, since it has the lower electron affinity.
(2) The total number of valence electrons is 26. There are 13 valence pairs.
(3) Place a single bond between each pair of atoms.

```
        Cl - N - Cl
            |
            Cl
```

(4) and **(5)** Ten pairs of electrons remain, but only nine are required to provide an octet around each chlorine. The remaining pair is used to serve as a lone pair that will complete the octet for the nitrogen atom.

```
        ..      ..      ..
      :Cl - N - Cl:
        ..    |     ..
            : C :
              ..
```

3. d. **(1)** Since only two atoms are present, there isn't really a central atom.
(2) The total number of valence electrons is the sum of the group numbers, that is, 6 + 4 = 10, and there are five valence pairs.
(3) Place a single bond between each pair of atoms.

```
            Ge - O
```

(4) and **(5)** The remaining four pairs are distributed around the two atoms as lone pairs.

```
          ..      ..
        : Ge : O :
```

(6) Both the germanium and the oxygen are deficient by one pair of electrons. To correct this, move one lone pair from each element into a bonding pair, making a triple bond between the two atoms.

```
        : Ge:::O :
```

Notice that the triple bond consists of one sigma and two pi bonds.

3. e. The chlorine is the central atom, and there are twenty valence electrons. First, insert single bonds between each pair of atoms.

```
          O - Cl - O
```

Of the remaining 16 electrons, only 12 are required around the "terminal" oxygens to complete their octets. The last two pairs of electrons are placed on the central chlorine atom, to complete its octet.

Thus the structure is

$$[\ddot{\text{:}}\ddot{\text{O}} - \ddot{\text{Cl}} - \ddot{\text{O}}\ddot{\text{:}}]^-$$

3. f. The nitrogen is the central atom, and there are sixteen valence electrons. First, insert single bonds between each pair of atoms.

O - N - O

Placing the remaining six electron pairs around the terminal oxygen atoms completes their octets, but the nitrogen needs two pairs of electrons.

$$[\ddot{\text{:}}\ddot{\text{O}} - \text{N} - \ddot{\text{O}}\ddot{\text{:}}]^+$$

Transfer one of the lone pairs from each oxygen to the bond between the nitrogen and the oxygen. This provides the electrons needed by the nitrogen.

$$[\ddot{\text{:}}\ddot{\text{O}} = \text{N} = \ddot{\text{O}}\ddot{\text{:}}]^+$$

This ion has two double bonds, that is, two sigma bonds and two pi bonds.

4. Except where indicated, the bond order of each bond is one. **[2]**   **[2]**

```
         O ✓                    H   O ✓
         ‖                      |   ‖
(a) H - O - C - H      (b) H - C - C - H
                               |
                               H
```

σ bonds   4   6
π bonds   1   1

5. a. The selenium is the central atom, and the number of valence electron is 24. Begin by placing one bond between each of the atoms.

```
O - Se - O
     |
     O
```

If the remaining 18 electrons (9 pairs) are distributed along the "terminal" oxygens, there are just enough electrons to provide an octet for each oxygen. The selenium lacks one pair of electrons, and normally we would transfer a pair from one of

170

the atoms bonded to the selenium. Since all three
oxygens are identical, which one forms the double
bond with selenium? This is the reason why
resonance forms are required, to make a double bond
equally possible along each of the selenium-oxygen
bonds.

```
   ..              ..              ..              ..              ..              ..
  :O - Se =  O:    ↔    :O - Se -  O:    ↔     O = Se -  O :
  ..    |          ..           ..    ||    ..           ..    |     ..
       :O:                           :O:                      :O:
        ..                                                     ..
```

b. Here, nitrogen is the central atom, and the
number of valence electron is also 24. Begin, as
usual, by placing one bond between each of the
atoms.

```
        O - N - O
            |
            O
```

As in the previous case, the remaining 18 electrons
(9 pairs) are just enough electrons to provide an
octet for each "terminal" oxygen. To provide the
required pair for the nitrogen, it's necessary to
transfer one pair from one of the atoms bonded to
the nitrogen. Since all three oxygens are
identical, this is another case where resonance
forms are required.

```
   ..              ..      -         ..              ..      -     ..              ..      -
 [ :O - N =  O:]    ↔   [:O - N -  O ]    ↔   [ O = N -  O:]
  ..    |          ..            ..    ||    ..           ..    |     ..
       :O:                            :O:                      :O:
        ..                                                      ..
```

6. a. Since the bond order is the same in all three
cases, and all three bonds involve nitrogen, the
only variable quantity is the halogen. The halogen
atoms increase in size in the order F<Cl<Br, and so
the order of the bond lengths is
        N-F < N-Cl < N-Br.

b. In this case the nitrogen atom and the bond
order are constant, so the bond length is
determined by the other bonded atoms, which
increase in size in the order N < P < As. Thus the
order of bond length is
        N-N < N-P < N-As

7. The balanced equation for the oxidation of one mole of $H_2(g)$ is

$$2 H_2(g) + O_2(g) \rightarrow 2 H_2O(g)$$

Set up a data table of bonds broken and bonds formed.

Bonds Broken	Bond Energy	Number Broken	Total Energy
H - H	436 kJ/mol	2 moles	872 kJ
O = O	498	1	498

Total Energy used to break these bonds = 1370 kJ

Bonds Formed	Bond Energy	Number formed	Total Energy
H - O	464 kJ/mol	4 moles	1856 kJ

Total Energy released forming these bonds = 1856 kJ

Net Energy Change = Energy Used - Energy Released

$$= 1370 \text{ kJ} - 1856 \text{ kJ}$$

Net Energy Change = -486 kJ

(Notice the change is negative since there is a net release of energy.)

Since this is the energy change for two moles of hydrogen gas, divide by two to obtain the enthalpy of combustion per mole of $H_2(g)$.

Standard Enthalpy of Combustion = -243 kJ/mol

8. C - I < C - N < C - O < C - F

9. MB_2E - b. bent (120°)
   MB_5E - e. square pyramidal
   MB_3E - d. trigonal pyramidal
   MB_3E_2 - c. "T-shaped"
   MB_2E_3 - f. linear
   MB_4E_2 - a. square planar

172

10.

formula	structural-pair shape	molecular or ionic shape
a. BeF_2	linear	linear
b. CCl_4	tetrahedron	tetrahedral
c. BCl_3	trigonal planar	trigonal planar
d. OF_2	tetrahedron	bent (~109.5°)
e. $BrCl_3$	trigonal bipyramid	"T-shaped"
f. BCl_2^-	trigonal planar	bent (~120°)

11. The electronegativity difference between the elements in this series of compounds increases in the order

increasing electronegativity difference →
CsI > CsBr> CsCl

and so the compounds become more ionic in the same order. Since increasing ionic character favors a higher melting point and decreased water solubility, the answers to the questions are

a.   increasing melting point →
CsI > CsBr> CsCl
b.   increasing solubility →
CsCl > CsBr > CsI

12. a. The Lewis structure for HCN is

$$H - C \equiv N :$$

Using the equation
formal charge = group number
- no. of unshared electrons
-1/2(no. of shared electrons)
For hydrogen
formal charge = 1 - 0 - 1/2(2) = 0
For carbon
formal charge = 4 - 0 - 1/2(8) = 0
For nitrogen
formal charge = 5 - 2 - 1/2(6) = 0

173

b. The Lewis structure for $POCl_3$ is

$$: O :$$
$$\overset{\cdot\cdot}{:} Cl - \overset{\overset{\parallel}{P}}{\underset{|}{}} - \overset{\cdot\cdot}{Cl} :$$
$$: Cl :$$

Using the same equation
    formal charge = group number
                  - no. of unshared electrons
                  -1/2(no. of shared electrons)
For oxygen
    formal charge = 6 - 4 - 1/2(4) = 0
For phosphorus
    formal charge = 5 - 0 - 1/2(10) = 0
For chlorine (notice that all three are identical)
    formal charge = 7 - 6 - 1/2(2) = 0

PRACTICE TEST (50 min.)

1. Draw Lewis structures for the following molecules or ions:

(a) O_2 (b) PH_3 (c) $AsCl_3$

(d) CCl_4 (e) SCl_2 (f) ClO_3^-

(g) CN^- (h) H_2CS

2. The following ions have two or more resonance structures. Show all of the resonance structures for each species.

(a) PO_3^- (b) NO_2^-

3. Determine the atom formal charges for each atom in the following species.

(a) OCS   (b) NF_3 (c) SCN^-

4. Arrange the following pairs of bonded atoms in terms of increasing bond length:

(a) C=C, C=O, and C=N
(b) C=S, C=Se, and C=O

174

5. What is the bond order of each chemical bond in the following molecules. In each case, indicate the number of sigma and the number of pi bonds present.

(a) $S = C = O$ (b)

```
            H       O   H
            |       ||  |
     H  -  C  -  O - C - C - H
            |           |
            H           H
```

(c) $H - C \equiv N$

6. Ethylene, C_2H_4, is one of the major starting materials used by the petrochemical industry. One way to produce it might be from ethane, C_2H_6, using the reaction

```
     H   H                          H         H
     |   |                           \       /
H -  C - C - H   →   H_2(g)  +        C = C    (g)
     |   |                           /       \
     H   H                          H         H
```

Using the table of bond energies provided, estimate the enthalpy change for this reaction. Does it seem likely that this will be a commercially practical way to produce ethylene? Why?

Bond	Bond Energy
C = C	611 kJ/mol
C - C	347
C - H	414
H - H	436

7. Identify each species that has a dipole moment and, where possible, indicate the direction of the net dipole by drawing an arrow with the arrowhead pointing to the negative end of the bond dipole. (Hint: Determine the structure with VSEPR Theory.)

a. O_2 b. OF_2 c. CCl_4

d. I_3^- e. PF_3 f. CS_2

8. Describe the central atom structural-pair shape and the molecular (or ionic) shape (including the <u>approximate</u> bond angle for bent molecules) for each of the following:

a. $SeCl_2$ b. $POCl_3$ c. H_2CS

d. SF_6 e. PF_5 f. NF_3

g. XeF_2 h. IF_4^+ i. $XeOF_4$

j. HPF_2 k. TeF_5^- l. SF_4

CONCEPT TEST ANSWERS

1. group number
2. ionic, covalent
3. ionic
4. electron affinity
5. one, carbon, nitrogen, oxygen, and sulfur
6. group number, 1/2
7. bond order
8. one, two, three
9. O-F < O-Cl < O-I
10. always endothermic
11. C-C < <u>C</u>-Cl < <u>Na</u>-Cl
12. zero
13. (a) octahedral   (b) trigonal planar
    (c) trigonal bipyramidal   (d) linear
14. transition metal atom or ion
15. dipole moment

PRACTICE TEST ANSWERS

1. (**L3**) a. $:\overset{..}{O} = \overset{..}{O}:$ b. $H - \overset{..}{P} - H$
                                                                $\qquad\qquad\qquad\quad |$
                                                                $\qquad\qquad\qquad\quad H$

c.
$$: \overset{..}{Cl} - \overset{..}{As} - \overset{..}{Cl} :$$
$$\qquad\quad |$$
$$\qquad :\overset{..}{\underset{..}{Cl}}:$$

d.
$$\qquad\quad :\overset{..}{Cl}:$$
$$\qquad\quad |$$
$$:\overset{..}{\underset{..}{Cl}} - C - \overset{..}{\underset{..}{Cl}}:$$
$$\qquad\quad |$$
$$\qquad :\overset{..}{\underset{..}{Cl}}:$$

e. :Cl - S̈ - Cl :   f. [:Ö - C̈l - Ö:]⁻

g.   [:C ≡ N:]⁻   h.   H - C - H
                                    ‖
                                   : S :

2. (**L4**)

a. [:O = P - Ö:]⁻ ↔ [:Ö - P = O:]⁻ ↔ [:Ö - P - Ö:]⁻
        | | ‖
       :O: :O: :O:

b.   [:O = N - Ö:]⁻   ↔   [:Ö - N = Ö:]⁻

3. (**L6**) a. sulfur 0, oxygen 0, carbon 0
   b. nitrogen 0, fluorine 0
   c. sulfur –1, carbon 0, nitrogen 0

4. (**L5**) a. C=O < C=N < C=C   b. C=O < C=S < C=Se

5. (**L3**)   All bond orders are one, except those
indicated, designated as two or three with bold
type and arrows.   [2]
         [2]   H   ↘ O   H   [3]
    ↙   ↘ | ‖ | ↘
a. S = C = O b. H - C - O - C - C - H c. H - C ≡ N
                     | |
                     H   H

σ bonds	2	10	2
π bonds	2	1	2

6. (**L5**) The calculated ΔH° = +128 kJ. Since this
reaction is highly endothermic, it's unlikely that
it will be useful commercially.

7. (**L8**) a. O = O The bond isn't polar, and the
   molecule doesn't have a dipole moment.
   b. OF_2 is a bent molecule, and the bonds are
   polar, so this molecule has a dipole moment.

c. CCl_4 is a tetrahedral molecule. The bonds are polar, but the symmetrical structure causes the molecule to be nonpolar.
d. I_3^- is a linear molecule, and the bonds are not polar, so this is also a nonpolar molecule.
e. The molecular structure of PF_3 is a trigonal pyramid, with a lone pair at one position in the tetrahedron of electron pairs. This will cause the molecule to be polar.
f. CS_2 is linear (like CO_2). The bonds are polar, but the molecule is not.

For the two polar molecules, OF_2 and PF_3, the directions of the dipole moments are as shown.

8. (L7)

formula	structural-pair shape	molecular or ionic shape
a. $SeCl_2$	tetrahedron	bent (109.5°)
b. $POCl_3$	tetrahedron	tetrahedron
c. H_2CS	trigonal planar	trigonal planar
d. SF_6	octahedron	octahedron
e. PF_5	trigonal bipyramid	trigonal bipyramid
f. NF_3	tetrahedron	trigonal pyramid
g. XeF_2	trigonal bipyramid	linear
h. IF_4^+	trigonal bipyramid	"see-saw" or distorted tetrahedron
i. $XeOF_4$	octahedron	square pyramid
j. HPF_2	tetrahedron	trigonal pyramid
k. TeF_5^-	octahedron	square pyramid
l. SF_4	trigonal bipyramid	distorted tetrahedron

178

CHAPTER 11
FURTHER CONCEPTS OF CHEMICAL BONDING:
ORBITAL HYBRIDIZATION, MOLECULAR ORBITALS
AND METALLIC BONDING

LEARNING GOALS:

1. Historically, the valence bond theory is one of the most important theories of chemical bonding. Understand how to apply this theory to a variety of simple molecules and identify the type of hybridization and the molecular structure in each case. This theory provides some new insights into how pi and sigma bonds form as well as the development of resonance structures. (Sec. 11.1)

2. Know how to draw molecular orbital electron configurations for simple homonuclear and heteronuclear molecules or ions, as well as understand the special perspective that MO Theory offers for understanding resonance. (Sec. 11.2)

3. The molecular orbital model is especially useful for explaining the bonding differences among insulators, semiconductors, and conductors. It also serves as the basis for the band theory of metallic bonding. Be able to explain each of these types of behavior in general terms using the molecular orbital theory. (Sec. 11.3)

IMPORTANT NEW TERMS:

antibonding molecular orbital (11.2)
band theory (11.3)
bonding molecular orbital (11.2)
conduction band (11.3)
Fermi level (11.3)
heteronuclear molecule (11.2)
homonuclear diatomic molecule (11.2)
insulator (11.3)
maximum overlap (11.1)
molecular orbital (MO) theory (11.2)

orbital hybridization (11.1)
semiconductor (11.3)
valence band (11.3)
valence bond (VB) theory (11.1)

CONCEPT TEST

1. When forming hybrid orbitals, the total number of hybrids formed must always equal _____ _____.

2. The molecular orbital theory is based on combining pure atomic orbitals from each atom in a molecule to form molecular orbitals that are _____ over the molecule.

3. The valence bond theory assumes that when a bond is formed between two atoms, the strongest possible sigma bond will be formed when the orbitals are arranged to give the _____.

4. For each type of hybrid orbital listed on the left below, indicate how many of each type of pure atomic orbital is combined to form the hybrids and the number of orbitals in a complete set of hybrids of that type.

hybrid type	no. of s orbitals	no. of p orbitals	no. of d orbitals	total hybrids
sp^2	_____	_____	_____	_____
sp^3d	_____	_____	_____	_____

5. Circle the elements that are <u>least</u> likely to exhibit expanded valence.

F   I   Se   N

6. Complete the table below by filling in the blanks to indicate the missing information.

hybrid orbital set	number of hybrid orbitals	geometry
sp^2	_____	_____
_____	4	_____
_____	_____	octahedral

7. In order for a pi bond to form, there must be _____ on the atom where the hybridization has occurred.

8. The first principle of molecular orbital theory is that the total number of molecular orbitals formed must always equal _____ _____.

9. Atomic orbitals are most effective in forming molecular orbitals when they combine with other orbitals of the same _____ and similar _____.

10. When all other factors are held constant, a. what is the relationship between bond order and bond length? b. between bond order and bond dissociation energy? a. _____ b. _____

11. Pi bonds do not occur unless the bonded atoms are already being joined by a _____ bond.

181

12. When a molecular orbital decreases the electron probability between the nuclei and causes the atomic nuclei to be attracted away from each other, the molecular orbital is said to be _____.

13. Another basic principle of molecular orbital theory is that the bonding molecular orbital is _____ (higher or lower) in energy than the parent orbitals, and the antibonding molecular orbital is _____ (higher or lower) in energy.

14. The elements called _____ are characterized by the fact that the valence band is only partly filled.

15. In a metal, the highest filled level at absolute zero is called the _____.

STUDY HINTS:

1. If you have difficulty determining how many atomic orbitals are actually hybridized, you may wish to use the VSEPR Theory to determine the structure. Then the corresponding type of hybridization becomes obvious if you have memorized the relationships between structure and type of hybridization.

2. Even though the molecular orbital theory may be new to you, don't forget that many of the rules you learned previously, such as Hund's rule and the Pauli principle, still hold true. These rules work here in much the same way that you learned when you did electronic configurations of atoms. For example, you can still depend upon the fact that each orbital, regardless of whether it is a pure

atomic orbital, a hybrid, or a molecular orbital, may contain a maximum of two electrons.

3. The answers in this chapter list the orbitals in the order that you would predict based on Figure 11.17. Remember that this energy order for the molecular orbitals is not strictly true for all molecules. It does, however, lead to the correct predictions of magnetism and bond order in the cases that you will encounter.

PRACTICE PROBLEMS

1. (**L1**) Tell what hybrid orbital set is used by the oxygen atom in each of the following molecules or ions:

a. O_3 b. CH_3-OH c. FOOF

_____   _____   _____

2. (**L1**) Tell what hybrid orbital set is used by the underlined atom in each of the following molecules or ions:

a. $\underline{N}F_3$ _____   b. $\underline{Si}F_4$ _____

c. $Cl_2\underline{C}=O$ _____   d. $\underline{Al}Cl_4^-$ _____

e. $F\underline{C}N$ _____   f. $\underline{P}Cl_3$ _____

g. $F\underline{N}NF$ _____   h. $O\underline{C}Se$ _____

3. (**L1**) Tell what hybrid orbital set is used by the underlined atom in each of the following molecules or ions:

a. $\underline{S}F_4$ _____   b. $\underline{As}Cl_5$ _____

c. $\underline{S}F_6$ _____   d. $\underline{Cl}F_4^+$ _____

e. $\underline{Al}Cl_6^{3-}$ _____   f. $\underline{Br}Cl_3$ _____

4. (**L2**) Determine the bond order and molecular orbital electron configuration for each of the following diatomic molecules or ions:

	Bond Order	Molecular Orbital Electron Configuration
a. Ne_2	_____	_____
b. Na_2	_____	_____
c. F_2^+	_____	_____
d. Al_2	_____	_____
e. Be_2^+	_____	_____
f. OF	_____	_____
g. O_2^{2-}	_____	_____

5. (**L2**) Which of the species in the previous question are diamagnetic?

6. (**L2**) Based on the bond order predicted by Molecular Orbital Theory, arrange the following species in order of increasing bond length: B_2^+, B_2, and B_2^-. What if you were asked to arrange these species in order of increasing dissociation energy?

7. (**L2**)   H - C = C - C = C - H
                      | | | |
                     H   H   H   H

Above is a compound called butadiene. What is the hybridization on each carbon atom in this molecule? Compare the length of the various carbon-carbon bonds in this molecule.

8. (**L3**) Consider three substances that will be designated A, B, and C. One of these substances is an insulator, one is a conductor and one is a semiconductor. If the band gap in A is 520 kJ/mole, the band gap in B is 105 kJ/mole, and the band gap in C is 7 kJ/mole, assign each of these substances to the most likely category.

184

1. a. sp^2 b. sp^3 c. sp^3

2a.

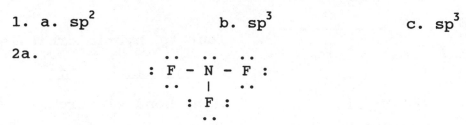

    The electron dot structure of NF_3 shows that there are four electron pairs (one lone pair and three bond pairs) that will be at the corners of a tetrahedron. The three bond pairs give the molecule a trigonal pyramidal structure.

    As shown in the diagram below, the s orbital and the three p orbitals on the nitrogen atom are hybridized to produce four orbitals, one of which contains a lone pair and the other three each contain one electron that is available for bonding.

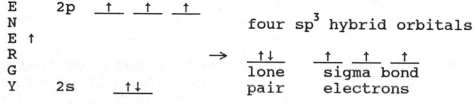

The hybridization on the nitrogen is sp^3.

2b.

: F :
      |
: F - Si - F :
      |
: F :

    The electron dot structure shows that there are four bond pairs at the corners of a tetrahedron, giving a tetrahedral structure.

    As shown in the diagram below, the s orbital and the three p orbitals on the silicon atom combine to produce four hybrids, each of which contains one electron that's available for bonding.

```
E     2p   _↑_ _↑_ _↑_
N                                        four sp³ hybrid orbitals
E ↑
R                        →          _↑_ _↑_ _↑_ _↑_
G
Y     2s      _↑_            sigma bond electrons
    isolated Si atom
```

The hybridization on the silicon is sp³.

2c.   : O :
                          ..   ‖   ..
                        : Cl - C - Cl :
                          ..   ..

The Lewis dot structure for this molecule indicates that there is a double bond from the carbon atom to the oxygen. Since this requires an unhybridized p orbital on the carbon atom, the hybridization must be as shown below.

```
      2p  _↑_ _↑_ _↑_                 _↑_
E
N                              unhybridized p orbital
E ↑                              used for pi bonding
R
G                       →          _↑_ _↑_ _↑_
Y
                              three sp² hybrid orbitals
      2s      _↑↓_             used for sigma bonds to
    isolated C atom           oxygen and chlorine
```

Thus the hybridization on the carbon atom is sp².

2d. The electron dot structure shows four sigma bonds connecting the aluminum to the four chlorines. The hybridization that matches this would be sp³.

2e. The Lewis dot structure is : F̈ - C ≡ N
                                    ..

The hybridization on the carbon must allow two pi bonds, so two p orbitals are not hybridized. The remaining p and s combine to form two sp hybrids.

186

2f. This molecule resembles NCl_3, with three sigma bonds and a lone pair. All three p orbitals and the s combine to form a set of four sp^3 hybrids, one of which is a lone pair.

2g. The Lewis dot structure is : F̈ - N̈ = N - F̈ :

There are two sigma bonds, one pi bond, and a lone pair on the nitrogen atom. This means that the hybridization on N must be sp^2.

2h. The Lewis dot structure is : Ö = C = S̈e :

There are two sigma and two pi bonds on the carbon, so only one p orbital participates in the hybridization. Thus the hybridization must be sp.

3a. The Lewis electron dot structure indicates that sulfur has an expanded set of valence electrons

```
            ..
          : F :
    ..      |   .  ..
  : F  -  S  '-  F :
    ..      |     ..
          : F :
            ..
```

This requires a set of sp^3d hybrids

```
       3d  ___ ___ ___ ___ ___      ___ ___ ___ ___
```

                                    4 unhybridized d
E   orbitals
N
E ↑ 3p ↑↓ ↑   ↑   → ↑↓ ↑   ↑   ↑   ↑
R
G   five sp^3d hybridized
Y   orbitals for four sigma
       3s ↑↓   bonds and one lone pair

    isolated S atom
The hybridization on the sulfur atom in SF_4 is sp^3d.

3b. The Lewis dot structure for $AsCl_5$ shows an expanded valence shell, with five sigma bonds to the five chlorines.

This means that the hybridization must be sp^3d.

3c. SF_6 has 24 electron pairs, which produces a Lewis dot diagram having six sigma bonds from the sulfur to the six fluorine atoms. To form these six bonds, the sulfur must have an expanded valence shell, using two of the d orbitals for hybridization.
   The hybridization must be sp^3d^2.

3d. The ClF_4^+ ion has 17 electron pairs. When they are distributed in a Lewis dot structure, the result is four sigma bonds from chlorine to fluorine and one lone pair.
   The hybridization is sp^3d.

3e. This molecule has 24 valence electron pairs, and there are six sigma bonds in the Lewis structure. To produce six sigma bonds on the aluminum, the valence shell must be expanded by the addition of two d orbitals.
   The hybridization is sp^3d^2.

3f. The $BrCl_3$ molecule has 14 valence electron pairs. In the Lewis structure, there are two lone pairs and three sigma bonding pairs on the bromine. This requires five hybrid orbitals.
   The hybridization on the bromine is sp^3d.

4.

		Bond Order	Molecular Orbital Electron Configuration
a.	Ne_2	0	$(\sigma_{2s})^2(\sigma^*_{2s})^2(\pi_{2p})^4(\sigma_{2p})^2(\pi^*_{2p})^4(\sigma^*_{2p})^2$
b.	Na_2	1	$(\sigma_{3s})^2$
c.	F_2^+	1 1/2	$(\sigma_{2s})^2(\sigma^*_{2s})^2(\pi_{2p})^4(\sigma_{2p})^2(\pi^*_{2p})^3$
d.	Al_2	1	$(\sigma_{3s})^2(\sigma^*_{3s})^2(\pi_{3p})^2$
e.	Be_2^+	1/2	$(\sigma_{2s})^2(\sigma^*_{2s})^1$
f.	OF	1 1/2	$(\sigma_{2s})^2(\sigma^*_{2s})^2(\pi_{2p})^4(\sigma_{2p})^2(\pi^*_{2p})^3$
g.	O_2^{2-}	1	$(\sigma_{2s})^2(\sigma^*_{2s})^2(\pi_{2p})^4(\sigma_{2p})^2(\pi^*_{2p})^4$

5. The diamagnetic molecules are Na_2 and O_2^{2-}. Ne_2 would be diamagnetic, but it doesn't exist.

6. In a list of similar molecules like this, the bond length will become larger as the bond order decreases. The expected order is B_2^- (1 1/2) < B_2 (1) < B_2^+ (1/2), with the bond order in parentheses after each species. The dissociation energy will become smaller as the bond order decreases, so the order of the molecules in terms of this quantity would be exactly the opposite of that for bond length.

7. Each carbon atom has the same hybridization, sp^2, and each of the carbon-carbon bonds must be the same length because what appears to be alternating double and single bonds is actually the same bond order in each case because of resonance.

8. The largest band gap is usually associated with an insulator, and so A is probably an insulator. The smallest band gap, for C, indicates a conductor, and the intermediate value for B, suggests that this material is a semiconductor.

PRACTICE TEST (45 min.)

1. Tell what hybrid orbital set is used by the underlined atom in each of the following molecules or ions:

a. $\underline{Te}Cl_4$

b. $\underline{O}F_2$

c. $H\underline{C}N$

d. $\underline{I}Cl_2^+$

e. IF_5

f. $\underline{Sn}Cl_2$

g. $\underline{I}F_4^+$

h. $\underline{Be}Cl_2$

i. $\underline{S}Cl_2$

j. $\underline{Si}F_6^{2-}$

k. $\underline{Sn}Cl_5^-$

l. $O\underline{C}S$

2. Tell what hybrid orbital set is used by the nitrogen atom in each of the following molecules or ions:

a. CH_3-NH_3

b. HCN

c. $NOBr$

3. Based on the bond order predicted by Molecular Orbital Theory, arrange the following species in order of (a) increasing bond length and (b) increasing dissociation energy: C_2^-, C_2^+, and C_2.

4. Predict the type of hybridization and the carbon-carbon bond angle for each carbon in the following molecule:

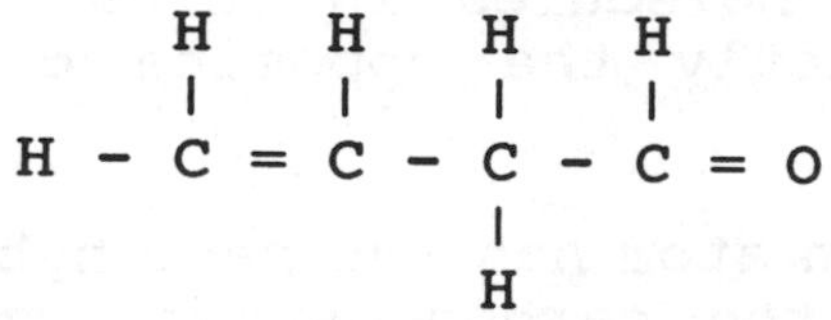

$$
\begin{array}{ccccccc}
& H & & H & & H & & H \\
& | & & | & & | & & | \\
H - & C & = & C & - & C & - & C & = O \\
& & & & & & | & \\
& & & & & & H &
\end{array}
$$

5.

$$
\begin{array}{l}
CH_3\text{-O} \quad S \leftarrow \qquad\qquad O \\
\qquad\quad \backslash \;\; \| \qquad\qquad \| \\
\qquad\qquad P \text{ -S-CH-C-O-CH}_2\text{-CH}_3 \\
\qquad\quad / \qquad\qquad | \\
CH_3\text{-O} \qquad\quad CH\text{-C-O-CH}_2\text{-CH}_3 \\
\qquad \uparrow \qquad\qquad\quad \| \\
\qquad\qquad\qquad\qquad\quad O
\end{array}
$$

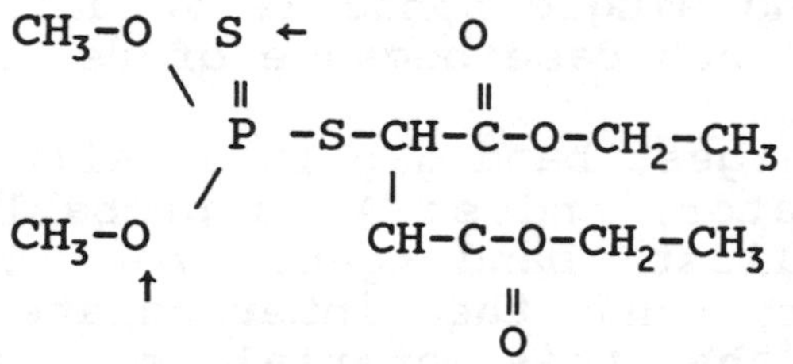

The structure of malathion, a commonly used pesticide, is shown above. Answer the question below based on the hybridization of this molecule.

a. How many sigma bonds are present in this molecule?   _____

b. How many pi bonds are present in this molecule?   _____

c. What is the hybridization on the oxygen marked with an arrow?   _____

d. What is the hybridization on the sulfur marked with an arrow?   _____

6. Thionyl tetrafluoride, OSF_4, is an interesting molecule because it is the first sulfur compound for which structure determinations have indicated that five groups are bonded to a central sulfur atom. (a) Predict the molecular structure and the

hybridization on the sulfur atom for this molecule. (b) Do you think it would be possible to synthesize a similar compound in which the sulfur were replaced by an oxygen atom? Do you think it would be possible to replace the sulfur with a selenium atom? Explain your answer in each case.

7. Determine the bond order and molecular orbital electron configuration for each of the following diatomic molecules or ions:

Bond Order	Molecular Orbital Electron Configuration
a. CF	
b. Cl_2^-	
c. C_2^+	
d. BeF	
e. NO^+	
f. B_2^+	

8. Which of the molecules and molecule-ions in the previous question will be paramagnetic?

9. Identify each of the following elements as a conductor, an insulator, or a semiconductor.

a. silicon
c. potassium

b. sulfur
d. chromium

10. Cyclohexane, C_6H_{12}, like benzene, is an organic compound which has a molecular structure based on a six-membered ring of carbon atoms. Draw the structure of this molecule and determine the type of hybridization on the carbon atoms. You may remember that the hybridization on the carbon atoms in a benzene molecule is sp^2, and the resulting molecule is planar. Do you think that the cyclohexane molecule will be planar?

11. a. The energy separation between the valence band and the conduction band is called the _____.
b. Substances that have a completely filled valence band are called _____.
c. Semiconductor materials that are created by adding small amounts of other elements (called dopants) are known as _____ semiconductors.
d. If the dopant has fewer valence electrons than the original substance, the resulting semiconductor is said to be a _____ semiconductor.

CONCEPT TEST ANSWERS

1. the number of pure atomic orbitals used in the combination.
2. delocalized   3. maximum overlap
4.

hybrid type	no. of s orbitals	no. of p orbitals	no. of d orbitals	total hybrids
sp^2	1	2	0	3
sp^3d	1	3	1	5

5. Since I and Se are the only elements in the group from the third or higher period, F and N are the least likely to exhibit expanded valence.
6.

hybrid orbital set	number of hybrid orbitals	geometry
sp^2	3	trigonal planar
sp^3	4	tetrahedral
sp^3d^2	6	octahedral

7. an unhybridized p atomic orbital
8. the number of atomic orbitals brought by the combining atoms.
9. type, energy
10. The higher the bond order, the shorter the bond length and the higher the bond dissociation energy.

11. sigma
12. antibonding
13. lower, higher
14. metals
15. Fermi level

PRACTICE TEST ANSWERS

1. (L1)
a. $\underline{Te}Cl_4$ — dsp^3
b. $\underline{O}F_2$ — sp^3
c. $H\underline{C}N$ — sp
d. $\underline{I}Cl_2^+$ — sp^3
e. $\underline{I}F_5$ — d^2sp^3
f. $\underline{Sn}Cl_2$ — sp^2
g. $\underline{I}F_4^+$ — dsp^3
h. $\underline{Be}Cl_2$ — sp
i. $\underline{S}Cl_2$ — sp^3
j. $\underline{Si}F_6^{2-}$ — d^2sp^3
k. $\underline{Sn}Cl_5^-$ — dsp^3
l. $O\underline{C}S$ — sp

2. (L1) a. sp^3 b. sp c. sp^2

3. (L2) In order of increasing bond length (bond order given in parentheses after each species):

$$C_2^- \ (2\ 1/2) < C_2 \ (2) < C_2^+ \ (1\ 1/2)$$

In order of increasing dissociation energy (bond order in parentheses):

$$C_2^+ \ (1\ 1/2) < C_2 \ (2) < C_2^- \ (2\ 1/2)$$

4. (L1) Reading from left to right, the hybridization on the carbon is sp^2, sp^2, sp^3, and sp^2. (Notice that two carbon-carbon single bonds separate the double bonds, so there will not be significant delocalization.) The approximate bond angles (in the same order) will be $120°$, $120°$, $109.5°$, and $109.5°$.

5. (L1)

CH₃-O   S ←sp²   O
        \ ‖   ‖
         P -S-CH-C-O-CH₂-CH₃
        / |
CH₃-O   CH-C-O-CH₂-CH₃
    ↑   ‖
   sp³   O

The molecule has 36 sigma bonds and 3 pi bonds.

6. **(L1)** a. The hybridization on the sulfur atom is dsp^3 and the structure is a trigonal bipyramid.
b. Oxygen is unlikely to form a compound of this type, since it can't form an expanded valence shell. Selenium can form an expanded valence shell, and so the selenium compound is possible.

7. **(L2)**

		Bond Order	Molecular Orbital Electron Configuration
a.	CF	2 1/2	$(\sigma_{2s})^2(\sigma^*_{2s})^2(\pi_{2p})^4(\sigma_{2p})^2(\pi^*_{2p})^1$
b.	Cl_2^-	1/2	$(\sigma_{3s})^2(\sigma^*_{3s})^2(\pi_{3p})^4(\sigma_{3p})^2(\pi^*_{3p})^4(\sigma^*_{3p})^1$
c.	C_2^+	1 1/2	$(\sigma_{2s})^2(\sigma^*_{2s})^2(\pi_{2p})^3$
d.	BeF	2 1/2	$(\sigma_{2s})^2(\sigma^*_{2s})^2(\pi_{2p})^4(\sigma_{2p})^1$
e.	NO^+	3	$(\sigma_{2s})^2(\sigma^*_{2s})^2(\pi_{2p})^4(\sigma_{2p})^2$
f.	B_2^+	1/2	$(\sigma_{2s})^2(\sigma^*_{2s})^2(\pi_{2p})^1$

8. **(L2)** All of these are paramagnetic except NO^+.

9. **(L3)** Based on their positions on the periodic table, silicon is a semiconductor, sulfur is an insulator, and both potassium and chromium are conductors.

10. **(L1)**

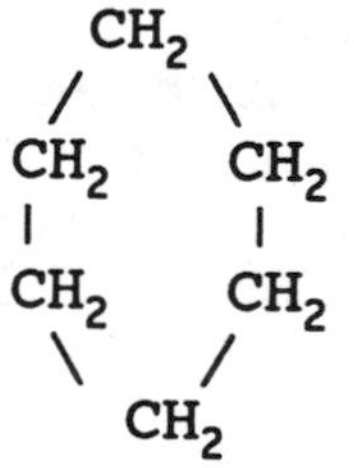

```
            CH2
          /     \
       CH2       CH2
        |         |
       CH2       CH2
          \     /
            CH2
```

The structure of cyclohexane is shown above. Each carbon atom displays sp^3 hybridization. Although the molecule appears to be planar, it is impossible to draw all of the carbon atoms in the same plane and maintain the 109.5° angles required by the sp^3 hybridization.

11. **(L3)** a. band gap   b. insulators,
        c. extrinsic   d. p-type

CHAPTER 12
GASES AND THEIR BEHAVIOR

LEARNING GOALS:

1. Understand how gas pressure is measured using a barometer or a manometer, and recognize the relationship among the various units of pressure, including millimeters of mercury, torr, atmospheres, pascals, and kilopascals. If necessary, review temperature conversions to insure that there will be no difficulty doing conversions involving degrees Celsius, degrees Fahrenheit, and Kelvins. (Sec. 12.1)

2. Understand the simple gas laws, including Boyle's law, Charles's law, and Avogadro's law. (Sec. 12.2)

3. Avogadro's hypothesis allows us to relate gas volumes to moles of gas and number of gas molecules. These relationships are often helpful when working problems. (Sec. 12.2)

4. The three gas laws mentioned in learning goal 2 can be combined to produce a useful relationship called the general gas law. This relationship can be used to solve many different types of gas law problems. (Sec. 12.3)

5. Another possible combination of the simple gas laws is called the ideal gas law. Compare this equation carefully with the general gas law to see when each will be most useful. Be especially sure to understand problems relating to gas densities or molar masses of gases, since the ideal gas law is particularly important for these calculations. (Sec. 12.3)

6. Stoichiometric calculations for chemical equations involving gases can be worked either by the methods learned in Chapter 5 or by means of Gay-Lussac's law, using gas volumes instead of moles. It's important to be able to use both

195

calculation methods. (Sec. 12.4)

7. Understand how to do gas law calculations for mixtures of gases based on Dalton's law of partial pressures and also be able to use mole fractions. (Sec. 12.5)

8. The kinetic molecular theory provides the basis for the currently accepted explanation of gas behavior. Understand the assumptions of this theory and also the important new terms related to the theory, such as average molecular kinetic energy, average molecular speed, and the distribution of molecular speeds. (Sec. 12.6)

9. Diffusion or effusion of gases is described in terms of Graham's law; know how to solve problems based on this relationship. (Sec. 12.7)

10. Understand why gas behavior doesn't follow the predictions of the ideal gas law and how to use the van der Waals equation to improve on these predictions. (12.8)

IMPORTANT NEW TERMS:

Avogadro's law (12.2)
barometer (12.1)
Boltzmann distribution curve (12.6)
Boyle's law (12.2)
Charles's law (12.2)
Dalton's law of partial pressures (12.5)
diffusion (12.7)
effusion (12.7)
Gay-Lussac's law (12.2)
general (or combined) gas law (12.3)
Graham's law of diffusion and effusion (12.7)
ideal gas (12.6)
ideal gas law (12.3)
intermolecular forces (12.8)
kinetic molecular theory of gases (12.6)
millimeter of mercury (torr) (12.1)
mole fraction (12.5)
non-ideal gas (12.8)
partial pressure (12.5)
pascal (Pa) (12.1)
pressure (12.1)
root-mean-square (rms) speed (12.6)

CONCEPT TEST

1. _____ is defined as the force exerted on an object divided by the area over which the force is exerted.

2. One standard atmosphere is equivalent to values of _____ mm of Hg or _____ kilopascals.

3. 720.0 mm Hg is equivalent to _____ kPa.

4. Boyle's law states that the volume of a confined gas is inversely proportional to the _____ exerted on the gas, if temperature is held constant.

5. Charles's law states that with pressure and quantity of gas constant, the volume of a gas will be directly proportional to the _____.

6. If the volume of a gas sample is graphed against the kelvin temperature while pressure is constant, a straight line results. By extrapolation to zero volume, this line results in a temperature value of _____ $^{\circ}$C, which is called absolute zero.

7. Avogadro stated that equal volumes of gases under the same conditions of temperature and pressure have equal numbers of _____.

8. Under conditions of standard temperature and pressure (STP), 1.0000 mole of any gas occupies a volume of _____ liters, a quantity called the _____.

9. The value of the gas constant, R, in the ideal gas equation is _____.
(Be sure to include the units!)

10. The total pressure exerted by a mixture of gases is the sum of the _____ of the individual gases in the mixture.

11. If the temperature on a confined gas sample is doubled, while the volume is held constant, what will happen to the pressure? _____

12. What does STP stand for? _____

13. For a given sample of gas molecules, the average kinetic energy depends only on the value of the _____.

14. The assumptions of the kinetic molecular theory are most likely to be incorrect for gases under which of the following combinations of conditions?

a. high temperature, high pressure
b. high temperature, low pressure
c. low temperature, high pressure
d. low temperature, low pressure

15. The values of the a and b constants in the van der Waals Equation are shown below for some typical gases.

gas	a	b
CCl_4	20.4 $L^2 \cdot atm/mol^2$	0.1383 L/mol
Kr	2.32	0.0398
O_2	1.36	0.0318

a. According to the values above, which of these three gases would you expect to display the smallest intermolecular attraction? _____

b. Which gas has molecules that you would expect to occupy the largest volume? _____

STUDY HINTS:

1. In order to work gas law problems, it is frequently necessary to organize many different numerical values. The best way to do this is by using a data table, and this procedure is followed throughout the textbook and this study guide. You may wish to use a modified version of the data table form shown here, but you will definitely want to use some type of data table to organize the information in these problems.

2. Neither Celsius nor Fahrenheit temperatures are useful for gas law calculations; *you must use kelvins*. Unfortunately it is easy to forget this when you become involved in a problem. To avoid this mistake, make it a point to always convert to kelvins as soon as possible and insert these values in the data table.

3. Students seem to be most likely to forget to convert to kelvins if the temperature is 273°C. It's not clear why this is true, but watch out for the situation and avoid this error.

4. Students sometimes forget the correct units that are required when values are inserted into the ideal gas law. If you memorize not just the numerical value but also the units of the gas constant, R, this should never be a problem.

5. You have probably noticed that the value of standard temperature is different for thermodynamics problems (25°C) than for gas law problems (0°C). Be sure to think about which problem type you are working so that you won't be confused by this difference.

PRACTICE PROBLEMS

1. (**L1**) A certain gas sample is measured at 720.0 mm Hg and 25°C; Calculate the corresponding temperature in kelvins and pressure in atmospheres.

2. (**L2**) A sample of oxygen gas has a volume of 128 milliliters at a pressure of 500.0 mm Hg. Calculate the volume this same sample would occupy at standard pressure.

3. (**L2**) A sample of nitrogen gas has a volume of 212 milliliters at a temperature of 57.0°C. Calculate the volume this same sample would occupy at standard temperature.

4. (**L3 or L5**) The density of an unknown gas measured at standard temperature and pressure is 1.96 grams/liter. What is the molar mass of this gas?

5. (**L3**) A high volume sampler used for air pollution work analyzes 1200 cubic meters of air. If this volume was measured at STP, how many molecules of air were sampled?

6. (L4) A certain mass of chlorine gas occupies a volume of 50.0 liters at a temperature of $273^{\circ}C$ and a pressure of 700.0 mm Hg. What volume will this same sample of gas occupy at STP?

7. (L5) How many moles of nitrogen gas are contained in a sample vial having a volume of 1.78 L if the temperature of the gas is $27.0^{\circ}C$ and the pressure is 750.0 mm Hg?

8. (L7) A mixture of gases consists of 56.0 grams of N_2, 16.0 grams of CH_4, methane, and 48.0 grams of O_2. If the total pressure of this mixture is 850.0 mm Hg, what is the mole fraction and partial pressure of each gas.

9. (**L5**) Calculate the pressure (in kilopascals) which will result if 2.5 grams of XeF_4 gas is introduced into an evacuated container which has a volume of 3.00 cubic decimeters and is kept at a constant temperature of $80.0^{\circ}C$. (In SI units,

$$R = 8.31 \ \frac{kPa \cdot dm^3}{mol \cdot K})$$

10. (**L9**) An unknown gas diffuses through a small opening at a rate of 23 mL/hour, but helium gas diffuses through the same opening at a rate of 92 mL/hour. What is the molar mass of the unknown gas?

11. (**L6**) The empirical formula of a certain hydrocarbon is CH_3, and when 0.500 mole of this hydrocarbon is completely combusted, 67.2 liters of carbon dioxide (measured at STP) is collected. What is the molecular formula of this hydrocarbon?

12. (**L7**) What is the pressure that results when 2.0 liters of hydrogen gas, measured at STP, is forced into a 2.0 liter container, which previously contained enough oxygen to completely fill the container under standard conditions. Assume that the temperature doesn't change when the gases are mixed.

13. (**L6**) At 383 K, methane, an organic compound, reacts with oxygen to form only gaseous products as indicated in the following equation:

$$CH_4(g) \quad + \quad O_2(g) \quad \rightarrow \quad CO(g) \quad + \quad H_2O(g)$$

If 0.010 mole of methane and 0.030 mole of oxygen are placed in a sealed container at an initial pressure of 1.00 atmospheres and the temperature is held constant during the reaction, calculate (a) the volume of the container, (b) the moles of carbon monoxide formed, and (c) the total pressure in the container at the end of the process.

PRACTICE PROBLEM ANSWERS

1. pressure = $\dfrac{720.0 \text{ mm Hg}}{760.0 \text{ mm Hg/atm}}$ = $\underline{0.947 \text{ atm}}$

    temperature = $25°C$ + 273 = $\underline{298 \text{ kelvin}}$

2. Remember that standard pressure is 760 mm Hg and use Boyle's law.

$$V = 128 \text{ mL} \times \dfrac{500.0 \text{ mm Hg}}{760.0 \text{ mm Hg}}$$

$$\underline{V = 84.2 \text{ mL}}$$

3. Use Charles's law. Don't forget to convert the temperature to kelvins.

$$V = 212 \text{ mL} \times \dfrac{273.2 \text{ K}}{(57.0 + 273.2) \text{ K}} = \underline{175 \text{ mL}}$$

4. First set up the data table

mass	1.96 g
volume	1.00 L
temp	273 K
pressure	1.00 atm
molar mass	?

The ideal gas law contains all of these variables.

$$PV = nRT = \dfrac{gRT}{M}$$

rearrange the equation to solve for molar mass

$$M = \dfrac{gRT}{PV}$$

$$M = \dfrac{1.96 \text{ g} \times 0.0821 \text{ L·atm/mol·K} \times 273 \text{ K}}{1.00 \text{ atm} \times 1.00 \text{ L}}$$

$$\underline{M = 43.9 \text{ g/mol}}$$

Notice that there is also an alternative method of solution, based on the observation that 1 mole of any gas at STP must occupy a volume of 22.414 L.

$$M = 1.96 \text{ g/L} \times 22.41 \text{ L/mol}$$

$$\underline{M = 43.9 \text{ g/mol}}$$

5. First you must convert 1200 m^3 into liters. Since a cubic meter is a cube 1 meter on a side, and one meter is 100 cm, the volume of a cubic meter in cubic centimeters is

$$V = 100 \text{ cm} \times 100 \text{ cm} \times 100 \text{ cm} = 1 \times 10^6 \text{ cm}^3$$

One cm^3 = one mL, so the volume is also 1×10^6 mL.

Thus, 1 m^3 is equivalent to 1000 L.

Using this conversion factor,

$$V = 1200 \text{ m}^3 \times 1000 \text{ L/m}^3 = 1.2 \times 10^6 \text{ L}$$

Since the volume is measured at STP, we can use the molar volume, 22.4 L/mol, to determine how many moles of air were sampled.

$$\text{moles} = \frac{1.2 \times 10^6 \text{ L}}{22.4 \text{ L/mol}} = 5.36 \times 10^4 \text{ mol}$$

Multiply the number of moles by Avogadro's number to obtain the number of molecules.

no. of molecules

$$= 5.36 \times 10^4 \text{ mol} \times 6.02 \times 10^{23} \text{ molecules/mol}$$

$$\underline{\text{no. of molecules} = 3.2 \times 10^{28} \text{ molecules}}$$

6. First, set up a data table.

	Initial conditions	Final conditions
Volume	50.0 mL	?
Pressure	700.0 mm Hg	760 mm Hg
Temperature	273 + 273 = 546 K	273 K
Moles	n_1	n_1

The data provided will fit well into the general gas law.

$$\frac{P_1V_1}{nT_1} = \frac{P_2V_2}{nT_2}$$

Now substitute the available data into the equation. Since the number of moles doesn't change, it will cancel out.

$$\frac{700.0 \text{ mm Hg} \times 50.0 \text{ mL}}{546 \text{ K}} = \frac{760.0 \text{ mm Hg} \times V}{273 \text{ K}}$$

Rearranging to isolate V, the unknown.

$$V = \frac{700.0 \text{ mm Hg} \times 50.0 \text{ mL} \times 273 \text{ K}}{546 \text{ K} \times 760.0 \text{ mm Hg}}$$

$$\underline{V = 23.0 \text{ mL}}$$

7. Again it is best to start by writing a data table. This time we will use a slightly different format to show that there is more than one way to organize a data table.

Volume $= 1.78$ L   Temperature $= 27 + 273 = 300$ K

Pressure $= \dfrac{750 \text{ mm Hg}}{760 \text{ mm Hg/atm}} = 0.987$ atm

Moles $= ?$

In this case, the data is appropriate for substitution into the ideal gas law, $PV = nRT$. Rearranging to isolate the unknown, n, produces

$$n = \frac{PV}{RT}$$

Substituting the values into this relationship,

$$n = \frac{0.987 \text{ atm} \times 1.78 \text{ L}}{0.08206 \text{ L·atm/mol·K} \times 300 \text{ K}}$$

$$\underline{n = 0.0713 \text{ moles}}$$

206

8. Since the grams of each gas is given, the obvious way to solve is to calculate the moles of each gas, then add to obtain the total moles, and finally determine mole fractions. Notice the data table looks different, but it still serves to organize the available information.

$$mol\ N_2 = \frac{56.0\ g}{28.02\ g/mol} = 1.999\ mol$$

$$mol\ CH_4 = \frac{16.0\ g}{16.04\ g/mol} = 0.998\ mole$$

$$mol\ O_2 = \frac{48.0\ g}{32.0\ g/mol} = 1.50\ mol$$

Total moles = mol N_2 + mol O_2 + mol CH_4

Total moles = 1.999 mol + 1.50 mol + 0.998 mol

Total moles = 4.497 mol

Now find the mole fraction of each gas.

Mole fraction N_2 = 1.999 mol/4.497 mol = __0.445 mol__

Mole fraction CH_4 = 0.998 mol/4.497 mol = __0.222 mol__

Mole fraction O_2 = 1.50 mol/4.497 mol = __0.334 mol__

Notice that the sum of all of the mole fractions equals 1.0, as is required by the definition of mole fraction.

To find the partial pressure of each gas, multiply the mole fraction times the total gas pressure.

Partial pressure N_2 = 0.445 x 850 mm Hg

= __378 mm Hg__

Partial pressure CH_4 = 0.222 x 850 mm Hg

= __189 mm Hg__

Partial pressure O_2 = 0.334 x 850 mm Hg

= <u>284 mm Hg</u>

9. As before, it is best to begin with a data table.

Temperature = 80.0 + 273.2 = 353.2 K
Pressure   = ?
Volume   = 3.00 dm³

For XeF_4, the molar mass is 207.3 g/mol

Thus,
moles of XeF_4 = 2.50 g / 207.3 g/mol = 0.0121 moles

All of the data needed for the ideal gas law is given, except that the volume is in SI units. Since the pressure is requested in kPa, and the appropriate value for R is also given, the problem can be solved by using the ideal gas law with SI units.

Rearranging PV = nRT to solve for P,

$$P = \frac{nRT}{V}$$

Substituting the values into this relationship,

$$P = \frac{0.0121 \text{ mol} \times 8.31 \text{ kPa}\cdot\text{dm}^3/\text{mol}\cdot\text{K} \times 353 \text{ K}}{3.00 \text{ dm}^3}$$

<u>P = 12 kPa</u>

10. Problems that concern diffusion or effusion are almost always based on Graham laws. To avoid confusion, be sure to set up a data table that defines which gas will be labelled 1 and which will be labelled 2.

	Molar mass	Rate
1. unknown	?	23 mL/hr
2. helium	4.00 g/mol	92 mL/hr

208

Graham law may be stated

$$\frac{rate_1}{rate_2} = \frac{\sqrt{M_2}}{\sqrt{M_1}}$$

inserting the values

$$\frac{23 \text{ mL/hr}}{92 \text{ mL/hr}} = \frac{\sqrt{4.00 \text{ g/mol}}}{\sqrt{M_1}}$$

$$M_1 = (8.00)^2 \text{ g/mol}$$

$$\underline{M_1 = 64 \text{ g/mol}}$$

11. The difficulty here is that we don't have a formula for the combustion; however, we do know that each mole of carbon in the original sample must produce one mole of carbon dioxide. It seems logical to begin by calculating the moles of carbon dioxide formed.

Since the gas is measured at STP, use the molar volume to determine the moles of CO_2.

$$\text{moles } CO_2 = \frac{67.2 \text{ L}}{22.41 \text{ L/mol}}$$

$$\text{moles } CO_2 = 2.999 \text{ moles}$$

Therefore, the combustion of 0.500 moles of hydrocarbon produces 2.999 moles of carbon dioxide. How many moles of carbon dioxide would be produced by combusting 1.00 mole of hydrocarbon?

mol CO_2 /mol of cpd = 2.999 mol CO_2 / 0.500 mol cpd

mol CO_2 /mol of cpd = 6.00 mol CO_2 /mol cpd

But each mole of CO_2 produced indicates one mole of carbon present in the compound. It must also be true that

mole C /mole of cpd = 6.00 mol C /mol cpd

The empirical formula tells us that there are three moles of hydrogen for each mole of carbon, so the molecular formula must be

$$\underline{C_6H_{18}}$$

12. Since the container is initially at standard pressure, the pressure on the oxygen alone must be 1.00 atm. When the gases are mixed, according to Dalton's law this becomes the partial pressure of the oxygen. Similarly, since the volume of the hydrogen is identical with that of the oxygen under similar conditions, the initial pressure of the hydrogen must also be 1.00 atm. If the temperature is left unchanged, the new total pressure must simply be the sum of the partial pressures, that is,

$$P_{total} = P_{O_2} + P_{H_2} = 1.00 \text{ atm} + 1.00 \text{ atm}$$

<u>And so the total pressure must be 2.00 atmospheres.</u>

13. a. A data table is useful here, but first use Dalton's law to find the total moles of gas.

total moles of gas = 0.010 mol + 0.030 mol
                   = 0.040 mol
temperature = 383 K
pressure   = 1.00 atm
volume   = ?

Substitute these values into the ideal gas law and solve for V

$$V = \frac{nRT}{P} = \frac{(0.040 \text{ mol})(0.08206 \text{ L atm/mol K})(383 \text{ K})}{1.00 \text{ atm}}$$

<u>V = 1.3 liters</u>

b. Begin by balancing the equation

$$2 \text{ CH}_4(g) + 3 \text{ O}_2(g) \rightarrow 2 \text{ CO}(g) + 4 \text{ H}_2\text{O}(g)$$

By inspection of the equation, it is apparent that 0.030 mole of oxygen will require 0.020 moles of methane for complete reaction. Since the amount of methane is insufficient, methane must be the limiting reagent.

$$\text{mol CO formed} = \text{mol CH}_4 \times \frac{2 \text{ mol CO}}{2 \text{ mol CH}_4}$$

<u>mol CO formed = 0.010 mole</u>

c. Let's review the moles of gas present at the end of the reaction:

    0.010 moles of CO formed

    0.0150 moles of oxygen didn't react

From the balanced equation, the moles of water formed will be twice the moles of carbon monoxide, so there will be 0.020 moles of water formed.

Total moles of gas

$$= 0.015 \text{ mol } O_2 + 0.010 \text{ mol CO} + 0.020 \text{ mol } H_2O$$

Total moles of gas = 0.045 moles

Now write the data table for the system when the reaction is completed.

    total moles of gas = 0.045 moles
    temperature = 383 K
    pressure   = ?
    volume   = 1.257 L

Substitute these values into the ideal gas law and solve for P

$$P = \frac{nRT}{V} = \frac{(0.045 \text{ mol})(0.0821 \text{ L atm/mol K})(383 \text{ K})}{1.26 \text{ L}}$$

<u>P = 1.1 atm</u>

PRACTICE TEST (50 minutes)

1. A sample of oxygen gas has a volume of 2.50 dm^3 at standard pressure. Calculate the volume (in dm^3) this same sample would occupy at 50.0 kPa.

2. Calculate the number of molecules of nitrogen in a pure sample of this gas having a volume of 0.210 liters at STP.

3. If the temperature and pressure are kept constant during the process, how many liters of titanium(IV) chloride will be produced when 10.0 liters of chlorine reacts with excess titanium according to the equation

$$Ti(s) \quad + \quad 2 \ Cl_2(g) \quad \rightarrow \quad TiCl_4(g)$$

4. A sample of nitrogen gas occupies a volume of 27.9 mL at a temperature of -80.0°C. The gas is heated until it occupies a volume of 85.5 mL. Assuming the pressure is constant, determine the new temperature of this sample in Celsius degrees.

5. A sample of oxygen gas having a volume of 56.0 milliliters is collected by water displacement at a temperature of 20°C and a pressure of 710.0 mm Hg. Determine the volume of this oxygen sample as a dry gas at STP. (The vapor pressure of water is 17.5 mm Hg at 20°C.)

6. If a certain gas sample has a volume of 30.0 cubic feet at a pressure of 500.0 mm Hg, what is the volume of this same sample in cubic feet at 2.0 atmospheres?

7. Calculate the molar mass of an unknown gas if at STP a sample of this gas has a density of 2.86 grams/liter.

8. Calculate the pressure in atmospheres which will result in a sealed container having a volume of 23.5 milliliters if 1.00 grams of BCl_3 gas is the only substance in the container and the temperature is 273°C.

9. Under the same conditions of temperature and pressure, methane gas (CH_4) diffuses through a porous barrier at a rate 2.3 times as fast as does a certain unknown gas. What is the molar mass of the unknown gas?

10. A certain gaseous compound, C_xH_y, contains only the elements carbon and hydrogen. When a sample of this compound is combusted, the only products are 5.38 L of carbon dioxide and 8.06 L of water, both measured at STP. (a) What is the empirical formula of the unknown compound? (b) If the volume of the original unknown gas was measured to be 2.69 L at STP, what is the molecular formula of the unknown compound?

11. A sample of unknown liquid is placed into a weighed, evacuated flask of known volume at a temperature high enough to vaporize all of the liquid. The temperature is held constant and the pressure in the flask is measured. The flask is weighed again to determine the mass of the unknown liquid. Using the data below obtained by this procedure, calculate the molar mass of this unknown liquid.

mass of the empty flask	35.364 grams
volume of the flask	35.0 mL
pressure in the flask	381 mm Hg
mass of the flask and the unknown	35.451 grams
temperature	$100.0\,^{\circ}C$

12. Hydrogen gas and oxygen gas are reacted to form water according to the balanced equation

$$2\ H_2(g)\ +\ O_2(g)\ \rightarrow\ 2\ H_2O(g)$$

If 20.0 liters of hydrogen is reacted with 30.0 liters of oxygen, calculate the total volume of gas at the end of the reaction. You may assume that the temperature is high enough that all of the water produced remains in the gas state, and that all gas volumes are measured at the same temperature and pressure.

CONCEPT TEST ANSWERS

1. pressure
2. 760.00 mm Hg, 101.325 kPa
3. 95.99 kPa   4. pressure
5. absolute or kelvin temperature
   (Not just temperature!)

6. $-273.15^{\circ}C$
7. molecules
8. 22.414 L, standard molar volume
9. 0.082057 L·atm/mol·K
10. partial pressures
11. it will double
12. standard temperature and pressure
13. temperature
14. c. low temperature, high pressure
15. a. O_2 b. CCl_4

PRACTICE TEST ANSWERS

1. (**L2**) 1.23 dm^3 2. (**L3**) 5.64x10^{21}
3. (**L6**) 5.00 liters   4. (**L2**) 319°C
5. (**L4**) 47.5 mL   6. (**L2**) 9.9 ft^3
7. (**L5**) 64.1 g/mol   8. (**L5**) 16.3 atm
9. (**L9**) 85 g/mol   10.(**L3**) CH_3, C_2H_6
11. (**L5**) 150 g/mol
12. (**L6**) 40.0 liters of gas

CHAPTER 13
INTERMOLECULAR FORCES, LIQUIDS, AND SOLIDS

LEARNING GOALS:

1. Intermolecular attractions play an important role in determining the properties of solids, liquids, and gases. These forces include ion-ion, ion-dipole, dipole-dipole, dipole-induced dipole, induced dipole-induced dipole interactions as well as hydrogen bonding. You should understand the way in which these forces arise and be aware of their relative strengths. (Sec. 13.2)

2. You should understand liquid behavior in general, in addition to specific properties, such as vaporization, vapor pressure, boiling, the critical point, surface tension, and viscosity. It is especially important to be able to calculate relationships between temperature and vapor pressure for pure liquids using the Clausius-Clapeyron equation. (Sec. 13.3)

$$\ln \left(\frac{P_2}{P_1} \right) = \frac{\Delta H_{vap}}{R} \left[\frac{1}{T_1} - \frac{1}{T_2} \right]$$

3. Be able to recognize the different types of solids, and the common cubic unit cells. Your knowledge of crystalline lattices should be sufficient to allow you to relate chemical formula to crystalline structure and to recognize how crystal packing characteristics help to determine the density of solids. (Sec. 13.4)

4. Since water is such an important compound for maintaining life on the Earth, you should be aware of the special properties of this compound and be able to explain these properties in terms of molecular structure and bonding theory. (Sec. 13.5)

5. A phase diagram shows how the various possible phases of a substance are affected by temperature and pressure. If you are given the phase diagram for a substance you should be able to use it to

discuss the effects of temperature and pressure changes on the behavior of that substance. You should also be able to draw a phase diagram when given appropriate information regarding a pure substance. (Sec. 13.6)

IMPORTANT NEW TERMS:

allotropes (13.4)
amorphous solid (13.4)
body-centered cubic (13.4)
boiling point (13.3)
capillary action (13.3)
Clausius-Clapeyron equation (13.3)
cooling curves (13.6)
critical point (13.3)
critical pressure (13.3)
critical temperature (13.3)
crystal lattice (13.4)
crystallization (13.4)
cubic close-packing (13.4)
dipole-dipole forces (13.2)
dipole-induced dipole forces (13.2)
dynamic equilibrium (13.3)
energy of hydration (13.2)
energy of solvation (13.2)
enthalpy of vaporization (13.3)
enthalpy of fusion (13.4)
enthalpy of vaporization (13.3)
evaporation (13.3)
face-centered cubic (fcc) (13.4)
heating curve (13.6)
hexagonal close-packing (13.4)
hydrated (13.2)
hydrogen bonding (13.2)
induced dipole-induced dipole forces (13.2)
intermolecular forces (13.2)
ion-dipole forces (13.2)
ion-ion forces (13.2)
kinetic molecular theory of liquids (13.3)
lattice energy (13.4)
London (or dispersion) forces (13.2)
melting point (13.4)
meniscus (13.3)
molecular solids (13.4)
network solids (13.4)

phase diagram (13.6)
polarizability (13.2)
polymer (13.4)
simple cubic (sc) (13.4)
solvation (13.2)
sublimation (13.4)
supercritical fluid (13.3)
surface tension (13.3)
triple point (13.6)
unit cell (13.4)
vapor pressure (13.3)
vaporization (13.3)
viscosity (13.3)
volatility (13.3)

CONCEPT TEST

1. Which of the three states of matter, gas, liquid, or solid, displays the smallest intermolecular forces? _____ In which state are the molecules arranged into the most regular array? _____

2. At normal temperatures and pressures, most of the elements are in the _____ state.

3. Coulomb's law indicates that the force of attraction between ions depends _____ on the ion charges and _____ on the distance between the ions. (Choose each answer from either directly or inversely.)

4. When a water molecule is attracted to an ion, the energy released in the process is called the

_____.

5. The negative slope of the solid/liquid equilibrium line on the phase diagram for water indicates that the solid form (ice) is _____

_____ (less dense than, more dense than, or of the same density as) the liquid form of water.

6. What are three elements most likely to cause hydrogen bonding when they are attached to hydrogen? _____

7. The degree to which the electron cloud of an atom or nonpolar molecule can be distorted by an external electric charge depends on the atom's or molecule's _____ .

8. The weakest of all intermolecular forces, often called dispersion forces or London forces, involves attraction between _____

9. When two opposing processes are occurring at exactly the same rate so that the effect of one cancels the effect of the other, the system is said to be in a state of _____ .

10. The curve of vapor pressure versus temperature on a phase diagram comes to a abrupt halt at the combination of critical temperature and pressure called the _____ .

11. The energy required to break through the surface of a certain liquid sample or to disrupt a drop of that liquid so that the material spreads out as a film is called _____ .

12. When a liquid is placed in a tube, the interaction among the adhesion of the liquid and the walls of the tube, the cohesion between the liquid molecules, and the force of gravity produces a characteristic concave or convex curve of the liquid surface called the _____ .

13. Different forms of the same element which can exist at the same temperature and pressure are called _____ .

14. When a solid is directly transformed into a gas without becoming a liquid, the process is called

_____.

15. On a phase diagram, the conditions of temperature and pressure at which three different phases can be in equilibrium is called the

_____.

16. Polyethylene and polypropylene are synthetic compounds used for many commonly encountered articles, including rugs, ropes, and bottles. These compounds are called _____ because they are formed by linking together a large number of monomeric chemical species in a large chain.

17. Each atom on the corner of a cubic unit cell will be shared by total of _____ unit cells; each atom in a face of a cubic unit cell will be shared by _____ unit cells, and each atom in the center of a cubic unit cell will be shared by _____ unit cells.

18. Substances can sublime only at temperatures and pressures that are below the _____.

PRACTICE PROBLEMS

1. (L1) in each of the following cases, determine what type of intermolecular force is involved and arrange the examples in order of increasing strength of attraction. a. NaF . . . H_2O, b. Rn . . . Rn, and c. NO . . . NO.

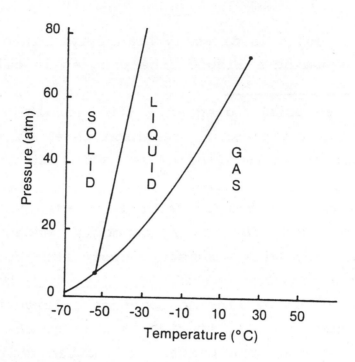

Figure 13.1 Rough Phase Diagram for CO_2

2. **(L5)** The above phase diagram is based on the following physical property data for carbon dioxide:

sublimation point $-78.5°C$ at 1.00 atm
triple point $-56.5°C$ at 5.11 atm
critical point $31.1°C$ at 72.9 atm

Use this information to answer the following questions regarding the behavior of carbon dioxide.

a. Which is more dense, solid CO_2, liquid CO_2, or do both states have the same density?   _____

b. What is the normal boiling point for carbon dioxide?   _____

c. Is CO_2 a gas, liquid, or solid at $-20°C$ and 65 atm?   _____

d. Approximately what pressure would be necessary to liquify CO_2 at $35°C$?   _____

3. (**L2**) How much heat will be required to completely vaporize 1.00 liter of butane at its boiling point. For liquid butane, the density is 0.601 g/mL, the molar mass is 58.13 g/mol, and the heat of vaporization is 24.3 kJ/mol.

4. (**L2**) Match the liquid property in the left hand column with the appropriate description from the right hand column by writing the appropriate letter in the space next to the property on the left.

_____ surface tension   a. causes water to rise in a tube

_____ critical point   b. energy required to overcome the attractive forces on the surface molecules

_____ capillary action   c. resistance to flow

_____ viscosity   d. highest temperature and pressure at which a gas can be liquified by application of pressure

5. (**L2**) Carbon disulfide is an unpleasant smelling liquid that has a vapor pressure of 298 mmHg at 20°C. If the atmosphere of a room having a volume of 30.0 liters is saturated with this vapor at 20°, how many grams of CS_2 are contained in the air?

221

6. (**L3**) Complete the table below by matching the type of solid from the left-hand column with the substance from the right-hand column that is an example of that type of solid.

_____ metallic solid   a. diamond
_____ ionic solid   b. I_2
_____ molecular solid   c. pure silver
_____ network solid   d. CsF

7. (**L2**) The vapor pressure of acetone, a common organic solvent is 283 mmHg at 30.0 degrees Celsius. If the enthalpy of vaporization for this compound is 31.4 kJ/mole, calculate the vapor pressure of acetone at 40.0 degrees Celsius.

8. (**L3**) The density of vanadium is 6.11 g/cm^3, and it crystallizes with a body-centered cubic unit cell. (a) Calculate the length (in nanometers) of a side of the unit cell for vanadium. (b) Based on this information, what is the radius of a vanadium atom?

PRACTICE PROBLEM SOLUTIONS

1. Sodium fluoride is ionic and water is polar, so the intermolecular forces involved will be ion-dipole. These are very strong, and since no other ionic attractions are present, this will be the strongest attraction.

Radon gas is nonpolar, so the only attractive forces will be induced dipole-induced dipole. These forces provide the weakest type of attraction.

Nitrogen oxide is a dipolar molecule, so the attraction is dipole-dipole, which is normally intermediate in strength between the previous two cases. Thus the order is

  b. Rn. . Rn < c. NO. . NO < a. NaF. . H_2O

2. a. solid CO_2 is more dense than liquid CO_2
   b. Liquid and solid carbon dioxide cannot be in
     equilibrium at 760 mmHg, and so this compound
     doesn't have a normal boiling point.
   c. liquid
   d. This temperature is above the critical
     temperature, and so carbon dioxide cannot be
     changed into a liquid by the application of
     pressure.

3. First, use the density equation to determine how many grams of butane are contained in 1.00 Liter.

$$\text{mass} = \text{density} \times \text{volume}$$

$$= 0.601 \text{ g/mL} \times 1.00 \text{ L} \times 1000 \text{ mL/L}$$

$$= 601 \text{ grams}$$

Next, divide by the molar mass to determine the number of moles.

$$\text{moles} = \frac{\text{grams of butane}}{\text{molar mass}}$$

$$= \frac{601 \text{ grams}}{58.13 \text{ g/mole}}$$

$$= 10.34 \text{ moles of butane}$$

Finally, determine the energy required.

$$q = 10.34 \text{ mol} \times 24.3 \text{ kJ/mol}$$

$$\underline{q = 251 \text{ kJ}}$$

4. surface tension — b. energy required to overcome the attractive forces on the surface molecules

critical point — d. highest temperature and pressure at which a gas can be liquified by application of pressure

capillary action — a. causes water to rise in a tube

viscosity — c. resistance to flow

5. Use the ideal gas law to calculate the amount of carbon disulfide that would be necessary to exert the observed vapor pressure.

$$\text{moles of } CS_2 = \frac{PV}{RT} = \frac{(298 \text{ mm/760 mm/atm})(30.0 \text{ L})}{(0.08206 \text{ L} \cdot \text{atm/mol} \cdot \text{K})(293.2 \text{ K})}$$

moles of CS_2 = 0.489 moles

mass of CS_2 = 0.489 moles x 76.14 g/mol

$\underline{\text{mass of } CS_2 = 37.2 \text{ grams}}$

6. metallic solid — c. pure silver
   ionic solid — d. CsF
   molecular solid — b. I_2
   network solid — a. diamond

7.
vapor pressure	283 mmHg	?
temperature (°C)	30.0	40.0
temperature (K)	303 K	313 K
enthalpy of vaporization	31.4 kJ/mole	

It appears that the Clausius-Clapeyron equation would be appropriate for this problem.

$$\ln\left(\frac{P_2}{P_1}\right) = \frac{H_{vap}}{R}\left[\frac{1}{T_1} - \frac{1}{T_2}\right]$$

Substituting the values provided into this equation

$$\ln\left(\frac{P_2}{283\ mm}\right) = \frac{31.4\ kJ/mol}{8.314\times10^{-3}\dfrac{L\cdot atm}{mol\cdot K}}\left[\frac{1}{303.2\ K} - \frac{1}{313.2\ K}\right]$$

$$\underline{\qquad P_2 \qquad} = \underline{\ 421\ mmHg\ }$$

8. a. First determine how many atoms of vanadium are present in a single unit cell. There are eight corner atoms, each shared by seven other unit cells, and one atom in the center that isn't shared with any other unit cell. Thus, the total is

```
8 corner atoms x 1/8 atom/site  =   1 atom
1 central atom (not shared)     = _ 1_
Total atoms per unit cell       =   2
```

This allows you to calculate the mass of a single unit cell:

$$\text{unit cell mass} = \frac{2\ atom/cell \times 50.94\ g/mole}{6.022\times10^{23}\ atom/mole}$$

unit cell mass = 1.6924×10^{-22} grams

Use the density to determine the volume of a unit cell.

$$\text{unit cell volume} = \frac{1.69\times10^{-22}\ grams}{6.11\ g/cm^3}$$

unit cell volume = $2.77\times10^{-23}\ cm^3$

Take the cube root of the volume to find the length of a side.

225

side of unit cell $=^3\sqrt{2.77\times10^{-23}}$ cm^3

$= 3.03\times10^{-8}$ cm

Convert to nanometers

Length of side (nm) $= \dfrac{3.03\times10^{-8} \text{ cm} \times 1\times10^{9} \text{ nm/m}}{100 \text{ cm/m}}$

Length of side = 0.303 nm

b. For a body-centered cubic cell, the contact between atoms occurs along the diagonal of the cube. The length of this diagonal is the square root of 3 times the length of the side of the cell.

Since that distance is equivalent to four times the radius of the atom

atomic radius $= \dfrac{s\sqrt{3}}{4} = 0.303$ nm x 0.4330

atomic radius = 0.131 nm

PRACTICE TEST (40 Minutes)

1. Arrange the following types of intermolecular attractions in order of increasing magnitude of the force exerted:
ion-dipole   induced dipole-induced dipole
hydrogen bonding   ion-ion

2. Using Coulomb's law, arrange the following ions in order of increasing energy of hydration: Ca^{2+}, Cs^+, and Ba^{2+}

3. Complete the table below by matching the type of solid from the left-hand column with the substance from the right-hand column that is an example of that type of solid.

_____ metallic solid   a. silicates
_____ ionic solid   b. an iron bar
_____ molecular solid   c. KI
_____ network solid   d. ice

226

4. The standard enthalpy of formation of liquid carbon tetrachloride is -135.44 kJ/mole, and the standard enthalpy of formation of gaseous carbon tetrachloride is -102.9 kJ/mole, what is the standard molar enthalpy of vaporization of CCl_4.

5. Carbon tetrachloride, CCl_4, is a toxic organic liquid that vaporizes rather readily. If an open container of carbon tetrachloride is placed in a sealed room at $20°C$ and allowed to come to equilibrium, it is observed that 18.4 grams of carbon tetrachloride are present in the 24,000 milliliters of air in the room. Based on this information, what is the vapor pressure of carbon tetrachloride in units of mmHg?

6. The vapor pressure of n-octane is 18.2 mmHg at 30.0 degrees Celsius and 118 mmHg at 70.0 degrees Celsius. Based on this information, calculate the enthalpy of vaporization for n-octane.

7. Metallic silver has a density of 10.5 g/cm^3 and crystallizes with a face-centered cubic unit cell. (a) Calculate the length of a side of the unit cell for silver. (b) Based on this information, what is the radius of a silver atom?

8. PHYSICAL PROPERTY DATA FOR CARBON MONOXIDE

normal melting point 68.09 K
normal boiling point 81.65 K
triple point   68.10 K at 115.4 mmHg

Using the information provided above, draw a rough phase diagram for carbon monoxide, and use your diagram to answer the questions below. You should label all axes, and indicate the stable state (gas, liquid, or solid) in each region of the diagram.

a. Is carbon monoxide a gas, a solid,
or a liquid at $71°K$ and 600 mmHg?   _____

b. Which is more dense, solid carbon
monoxide, liquid carbon monoxide,
or do both states have the same density?_____

227

c. Will carbon monoxide sublime at a
pressure of 1 atm? If not, indicate
a pressure at which it will sublime. _____

d. If the critical point for CO is at
132.9K and 34.5 atmospheres, approximately
what pressure would be necessary to
liquify CO at a pressure of 35 atm? _____

CONCEPT TEST ANSWERS

1. gas, solid
2. solid
3. directly, inversely
4. heat (or energy) of hydration
5. less dense
6. Nitrogen, oxygen, and fluorine
7. polarizability
8. between two induced dipoles
9. dynamic equilibrium
10. critical point
11. surface tension
12. meniscus
13. allotropes
14. sublimation
15. triple point
16. polymers
17. eight, two, and one. Notice that the question
    asks for the total number of unit cells sharing
    the atom.
18. triple point

PRACTICE TEST ANSWERS

1. (L1) induced dipole-induced dipole
   < hydrogen bonding < ion-dipole < ion-ion
2. (L1) Cs^+ < Ba^{2+} < Ca^{2+}
3. (L3) metallic solid   b. an iron bar
       ionic solid   c. KI
       molecular solid   d. ice
       network solid   a. silicate
4. (L2) +32.5 kJ/mole
5. (L2) 91 mmHg
6. (L2) 40.4 kJ
7. (L3) a. 0.409 nanometers b. 0.145 nm

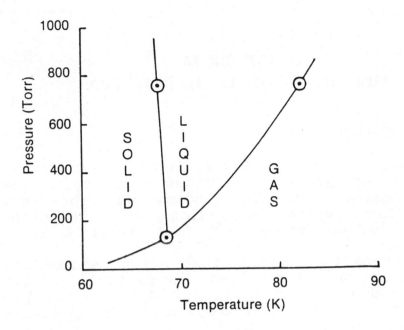

Figure 13.2 Approximate Phase Diagram for
Carbon Monoxide

8. **(L5)** The phase diagram is shown above.
   a. gas
   b. liquid
   c. no, below 115.4 mmHg
   d. no, this is above the critical temperature.

CHAPTER 14
SOLUTIONS AND THEIR BEHAVIOR

LEARNING GOALS:

1. If necessary, you should review the definitions of solute, solvent, solution, and molarity. This will prepare you to understand the new concentration units, including mole fraction, molality, and weight percent. (Sec. 14.1)

2. Understand the factors that affect solubility for various types of solutions and be familiar with the use of Henry's law and le Chatelier's principle to predict the effects of temperature and pressure on solubility. (Sec. 14.2)

3. You should understand how a nonvolatile solute can affect solvent vapor pressure and also be able to use Raoult's law to calculate the magnitude of this effect. Raoult's law can also be used to determine the molar mass of nonvolatile solvents by measuring the decreased vapor pressure of a solvent. (Sec. 14.3)

4. Freezing point depression and boiling point elevation due to the presence of dissolved solutes are frequently encountered applications of the theories discussed in this chapter. You should be able to work problems based on freezing point and boiling point change for both nonionizing solutes as well as solutes that ionize to produce more than one mole of particles per mole of solute. (Sec. 14.3)

5. Recognize that osmotic pressure is similar to the colligative properties studied previously and be able to do calculations based on the equation that relates osmotic pressure and molarity. (Sec. 14.3)

$$\Pi = MRT$$

6. Colloidal dispersions are commonly encountered in everyday life. Recognize how colloids differ from true solutions, and be able to identify some common emulsifying agents, including soaps and detergents. (Sec. 14.4)

IMPORTANT NEW TERMS:

boiling point elevation (14.3)
colligative properties (14.3)
colloidal dispersion (14.4)
detergent (14.4)
emulsion (14.4)
enthalpy of solution (14.2)
freezing point depression (14.3)
hard water (14.4)
Henry's law (14.2)
ideal solution (14.3)
immiscible (14.2)
Le Chatelier's principle (14.2)
miscible (14.2)
molality (14.1)
osmosis (14.3)
osmotic pressure (14.3)
Raoult's law (14.3)
saturated (14.2)
semipermeable membrane (14.3)
soap (14.4)
supersaturated (14.2)
surfactant (14.4)
suspension (14.4)
unsaturated (14.2)
weight percent (14.1)

CONCEPT TEST

1. Properties that depend only on the number of solute particles per solvent molecule and not on the nature of the solute or solvent are called _____ properties.

2. The sum of all of the mole fractions of all components in a solution must always equal a value of _____.

3. _____ is the term used to describe the maximum amount of material that will dissolve in a given amount of solvent at a given temperature to produce a stable solution.

4. When a solution temporarily contains more solute than the saturation amount, the solution is said to be _____.

5. If two liquids mix to an appreciable degree to form a solution they are said to be _____.

6. The general rule of thumb for predicting solubilities is that like dissolves like. For instance, polar solvents are most likely to dissolve _____ solutes, and nonpolar solutes are most likely to dissolve _____ solutes.

7. The solubility of gases in liquid solvents always _____ (choose from increases or decreases) when the temperature increases.

8. The solubility of most solids will increase when the temperature increases because this is usually an _____ (choose from exothermic or endothermic) process.

9. The lowering of the vapor pressure of a solution compared to the vapor pressure of the pure solvent is proportional to the _____ of the solute.

10. The boiling point of a solution is _____ than that of the pure solvent, whereas the freezing point of a solution is _____ than that of the pure solvent. (Choose your answers from higher or lower.)

11. As a solution freezes, solvent molecules are removed from the liquid phase and the solution becomes more concentrated, causing the freezing point of the remaining solution to _____.

12. In normal osmosis, molecules spontaneously move across a membrane to a solution where the concentration of solvent molecules is _____. (Choose from higher or lower.)

13. _____ consist of very finely divided particles that are too small to settle out of suspension in a solvent but too large to be considered a true solution.

14. Colloidal dispersions of one liquid in another, such as oil in water, are called _____.

15. A thin sheet of material that will allow only certain types of molecules to pass through it is called a(n) _____ membrane.

STUDY HINTS:

1. At first glance, molarity and molality seem to be very similar, but notice that molarity involves the <u>volume</u> (liters) of <u>solution</u> and molality involves the <u>mass</u> (kilograms) of <u>solvent</u>. Learn these differences carefully, so that they won't be overlooked in the pressure of an examination.

2. Remember that concentration units are extensive properties, that is, the same concentration value can be true regardless of the amount of solution present. Since the concentrations are true regardless of how much solution we have, in some problems it will be necessary to assume a mass or volume of solution. This is permitted as long as no values are given that limit the quantity of solution and as long as only one such assumption regarding the quantity of solution is made in a given problem.

233

3. Even after you are convinced that it is mathematically correct to assume some amount of solvent, solute, or solution in some of these problems, you may still feel a little confused about which quantity is the best to choose. Unfortunately there is no hard and fast rule you can memorize, but in most cases it is simplest to assume some amount of the substance in the denominator of the definition of the concentration unit that has a known value. Thus if you know the molality, assume a mass of solvent; if you know molarity, assume a volume of solution, and if you know the weight percent, assume a mass of solution. This plan of attack should provide you with some guidance until you become more familiar with this type of problem.

PRACTICE PROBLEMS

1. (L2) The rule of thumb that "like dissolves like" often provides practical help when preparing solutions. Suppose that only two solvents are available, toluene (a relatively nonpolar organic liquid) and water. For each of the following substances, predict which of these two solvents will be most likely to dissolve the material.

a. LiCl
c. benzene

b. acetic acid
d. methane, CH_4

2. (L1) The density of a 4.10 molal solution of NaCl is 1.120 grams/mL. Use this information to calculate (a) the weight percent and (b) the molarity of this solution.

3. (**L4**) a. Suppose that you have 1.00 mole of each of the following compounds: $BaCl_2$, $CsCl$, and $C_2H_4(OH)_2$ (ethylene glycol). Which will cause the greatest decrease in the freezing point of a 10.0 liter sample of water? Why is this true?

b. Suppose that you have 1.0×10^2 grams of each of the above compounds. Now which sample will cause the greatest decrease in the freezing point of the water? Explain this.

4. (**L3**) Consider a solution consisting of 50.0 grams of benzene (78.12 g/mole) and 50.0 grams of n-octane (molar mass = 114.2 g/mole). At $20^{\circ}C$ the vapor pressure of pure benzene is 95.2 mmHg, and the vapor pressure of pure n-octane is 14.1 mmHg. Use Raoult's Law to calculate the equilibrium vapor pressures of benzene and n-octane over this solution.

5. (**L5**) If 95.2 grams of a protein sample is dissolved in enough water to produce 1950 milliliters of solution, the resulting osmotic pressure is 0.0675 atmospheres at 25 degrees Celsius. Calculate the molar mass of this protein.

6. (L4) Calculate the freezing point of the solution in an automobile radiator that is 30.0% by weight ethylene glycol, $C_2H_4(OH)_2$. (Remember that the freezing point depression constant for water is $-1.86°C/m$ for water.)

7. (L6) A number of different terms are used to describe colloidal dispersions. For each term given, identify what is special about that type of colloidal dispersion, or if the term does not refer to a colloid, explain how it differs from a colloid.

a. emulsion

b. suspension

c. sol

d. gel

8. (L2) In some areas the well water has an odor like rotten eggs due to dissolved hydrogen sulfide gas. If the value of the Henry's law constant for H_2S is 2.6×10^{-4} molal/mmHg at $25°C$, and the concentration of hydrogen sulfide in the water was 0.050 molal, what was the pressure in atmospheres of the hydrogen sulfide gas when it was dissolved underground?

PRACTICE PROBLEM SOLUTIONS

1. Lithium chloride and acetic acid are both polar solutes, and so they are more likely to be soluble in a polar solvent like water. Benzene and methane are relatively nonpolar, and so are expected to be more soluble in toluene, a nonpolar solvent.

2. Since we are given the molal concentration, it will probably be best if we assume a given mass of solvent. This will be most directly useful with the definition of molality. We will begin by assuming we have 1.00 kilograms of water.

a. First find the total moles of NaCl
        moles of solute = kg of solvent x molality
                      = 1.00 kg x 4.10 m
                      = 4.10 moles

Next find the mass of NaCl
        mass of NaCl   = 4.10 moles x 58.44 g/mole
                      = 239.6 grams

Total mass of solution
                = mass of NaCl + mass of water
                = 239.6 g + 1000. g
                = 1239.6 g

   Percent NaCl = $\dfrac{239.6 \text{ g x } 100}{1239.6 \text{ g}}$

   Percent NaCl = 19.3 %

b. In order to use the molarity equation, you must know the volume of the solution. This can be determined from the density and the total mass of the solution.

   Density = mass/volume
   Volume of solution = $\dfrac{\text{mass}}{\text{density}}$ = $\dfrac{1239.6 \text{ g}}{1.120 \text{ g/mL}}$

   Volume of solution = 1106.8 mL

237

Now you know the volume of solution and the moles of solute, and so you can use the molarity equation.

$$\text{molarity} = \frac{\text{moles of solute}}{\text{liters of solution}}$$

$$\text{molarity} = \frac{4.10 \text{ moles}}{1.1068 \text{ liters}}$$

molarity = 3.70 M

3. a. Barium chloride and cesium chloride are both ionic compounds; assuming that the dissociation is essentially complete, one mole of barium chloride will produce three moles of ions, and one mole of cesium chloride will produce two moles of ions. The ethylene glycol doesn't ionize, and so it will produce only one mole of solute particles. Since freezing point depression depends on moles of solute particles, the barium chloride will cause the greatest change in the freezing point.

b. The molar masses of the three compounds are not the same, so 100 grams does not represent the same number of moles of each. The explanation is probably clearest if based on a data table, such as that below.

compound	molar mass	moles of cpd per 100 g	moles of ions per 100 g
$C_2H_4(OH)_2$	62.1 g/mole	1.61 moles	1.6 moles
CsCl	168	0.595	1.2
$BaCl_2$	208	0.481	1.4

Under these conditions, ethylene glycol produces the greatest number of ions in solution, and so it will cause the greatest freezing point depression.

4. First, calculate the moles of each compound:

$$\text{moles of benzene} = \frac{50.0 \text{ g}}{78.12 \text{ g/mole}} = 0.6400 \text{ moles}$$

$$\text{moles of octane} = \frac{50.0 \text{ g}}{114.2 \text{ g/mole}} = 0.4378 \text{ moles}$$

Calculate the mole fraction of each component:
mole fraction of benzene = X_b

$$= \frac{0.6400 \text{ mol}}{0.6400 \text{ mol} + 0.4378 \text{ mol}}$$

$$= 0.5938$$

mole fraction of octane $= X_o$

$$= \frac{0.4378 \text{ mol}}{0.6400 \text{ mol} + 0.4378 \text{ mol}}$$

$$= 0.4062$$

Finally, you are ready to determine the vapor pressure of each component of the mixture.

$P(\text{benzene}) = X_b P^o = 0.5938 \times (95.2 \text{ mmHg})$

$$= \underline{56.5 \text{ mmHg}}$$

$P(\text{octane}) = X_o P^o = 0.407 \times (14.1 \text{ mmHg})$

$$= \underline{5.72 \text{ mmHg}}$$

5. Rearranging the defining equation for osmotic pressure

$$M = \frac{\Pi}{RT} = \frac{0.0675 \text{ atm}}{(0.08206 \text{ L·atm/mol·K})(298 \text{ K})}$$

$$M = 0.002760 \text{ mole/L}$$

Using the definition of molarity

$$molarity = \frac{mol\ of\ solute}{Liters\ of\ solution}$$

$$molarity = \frac{g\ of\ solute/molar\ mass}{Liters\ of\ solution}$$

$$molar\ mass = \frac{95.2\ g}{(0.00276\ M)(1.95\ L)}$$

molar mass = <u>17,700 g/mole</u>

6. We must assume some quantity of solvent, solute, or solution, and as usual, it is simplest to assume some amount of the substance in the denominator of the definition of the concentration unit that is provided, that is, weight percent. Thus, we will assume 1000 grams of solution.

It is apparent then that the solution will consist of 300.0 grams of ethylene glycol and 700.0 grams of water.

$$moles\ of\ solute = \frac{grams\ of\ solute}{molar\ mass}$$

$$moles\ of\ solute = \frac{300.0\ g}{62.07\ g/mol} = 4.833\ mole$$

The kilograms of solvent is 0.7000 kg, and these values can be substituted into the defining equation for molality.

$$molality = \frac{moles\ of\ solute}{kilograms\ of\ solvent}$$

$$molality = \frac{4.833\ moles}{0.7000\ kg} = 6.905\ m$$

Finally, this value can be substituted into the freezing point depression relationship

$$\Delta T = K_f m = -1.86 \ ^\circ C/m \times 6.905 \ m$$

$$\underline{\Delta T = -12.8 ^\circ C}$$

Freezing point of the radiator solution = -12.8°C.

7. a. An emulsion is a colloidal dispersion of one liquid in another.
b. A suspension is _not_ a colloidal dispersion, since the particles are so large that it will settle out fairly rapidly.
c. A sol is a general term for a colloidal dispersion. The term may be modified by addition of a prefix, for example, an aerosol is a colloid that's dispersed in a gas.
d. A gel is a colloidal dispersion that has solidified and so resists flow.

8. Using Henry's law

$$molality = kP$$

rearranging

$$P(mmHg) = \frac{m}{k} = \frac{0.050 \ m}{2.6 \times 10^{-4} \ m/mmHg}$$

$$P(mmHg) = 192 \ mmHg$$

$$P(atm) = 192mmHg/760 \ mm/atm$$

$$\underline{P(atm) = 0.25 \ atm}$$

PRACTICE TEST (45 min)

1. Calculate the osmotic pressure of a solution prepared by dissolving 42.8 grams of sucrose (molar mass = 342.3 g/mole) in 250.0 milliliters of water at a temperature of 17.0°C.

2. A solution is prepared that contains 147.2 grams of H_2SO_4 in exactly 1.00 liters of solution. The density of the solution is 1.090 g/mL. Calculate

the (a) molarity, (b) molality, (c) weight percent, and (d) mole fraction of H_2SO_4 in this solution.

3. The freezing point depression constant for benzene is $-4.90^\circ C/m$, and the freezing point of pure benzene is $5.5^\circ C$. What is the freezing point of the solution that results when 8.15 grams of the organic compound having the formula $C_2H_2Cl_4$ is dissolved in 151 grams of benzene? Assume that the organic compound doesn't dissociate in benzene.

4. Hexane is an organic compound having a molar mass of 86.2 g/mole and a vapor pressure of 153 mmHg at $25^\circ C$. When 25.0 grams of an unknown organic liquid is mixed with 50.0 grams of hexane, the vapor pressure of hexane over the resulting solution is lowered to 107 mmHg. Assuming ideal behavior, what is the molar mass of the unknown liquid?

5. Carbonated soft drinks are produced by saturating water with carbon dioxide under pressure. In one such operation, the pressure of the carbon dioxide is 3.5 atmospheres at $25^\circ C$. What is the concentration of the CO_2 in the soft drink under these conditions? (At $25^\circ C$ the Henry's law constant $= 4.44 \times 10^{-5}$ m/mmHg for CO2.)

6. What is the boiling point of a solution prepared by dissolving 120.0 grams of NaCl in 500.0 grams of water? You may assume that the NaCl is 100% dissociated. ($K_b = 0.52^\circ C/m$ for water)

7. Calculate the vapor pressure of water over a 0.50 m solution of K_2SO_4 at $50^\circ C$. The vapor pressure of water at $50^\circ C$ is 92.5 mmHg. You may assume the potassium sulfate is 100% dissociated.

CONCEPT TEST ANSWERS

1. colligative
3. solubility
5. miscible
7. decreases
9. mole fraction
11. decrease further

2. one
4. supersaturated
6. polar, nonpolar
8. endothermic
10. higher, lower
12. lower

13. colloidal dispersions (or colloids)
14. emulsions
15. semipermeable membrane

PRACTICE TEST ANSWERS

1. **(L5)** 11.9 atm
2. **(L1)** a. 1.50 M   b. 1.59 m   c. 13.5%
   d. mole fraction = 0.0278
3. **(L4)** $3.92^{\circ}C$
4. **(L3)** 1.00×10^{2} g/mol
5. **(L2)** molality CO_2 = 0.118 m
6. **(L4)** $104.3^{\circ}C$
7. **(L3)** mole fraction of water = 0.974,
   vapor pressure = 90.1 mmHg

CHAPTER 15
CHEMICAL KINETICS: THE RATES AND
MECHANISMS OF CHEMICAL REACTIONS

LEARNING GOALS:

1. One of the fundamental pieces of information for kinetic studies is the rate of a chemical reaction. It's important to be aware of the factors that affect the speed of a chemical process, know something about how reaction rates are measured, and be able to distinguish between initial rate and instantaneous rate. (Sec. 15.2)

2. One way to express the effects of reactant concentrations on the rate of a reaction is by means of a rate expression. Know how to determine the rate expression from data that shows how the initial rate varies with changes in the concentration of one of the reactants (Sec. 15.3)

3. It's also important to understand how the integrated form of the first order rate expression and the half-life of a reaction can be a measure of reaction rate. Be able to use the equation

$$\ln \frac{[A]}{[A]_o} = -kt$$

to perform rate calculations on simple systems and know how to determine the rate expression by graphing the time against the natural logarithm of the concentration and the reciprocal of the concentration to determine in which case a linear graph results. (Sec. 15.4)

4. The transition state theory and the collision theory provide an explanation of how chemical reactions occur. Understanding these theories is an important theoretical background for all of the discussions in this chapter. (Sec. 15.5)

5. Be able to use the Arrhenius equation

$$\ln \frac{k_2}{k_1} = -\frac{E^*}{R} \left[\frac{1}{T_2} - \frac{1}{T_1} \right]$$

to predict the effects of temperature on reaction rate. This will require that the equation be solved either directly or by means of a graph.

6. Understand how elementary steps are combined to develop a reaction mechanism and the meaning of the terms reaction intermediate and rate determining step. Although students working with kinetics for the first time will not yet have enough experience to suggest reaction mechanisms, it should be possible to examine a proposed mechanism and determine if it agrees with the available experimental data. (Sec. 15.6)

IMPORTANT NEW TERMS:

activated complex (15.5)
activation energy (15.5)
Arrhenius equation (15.5)
bimolecular (15.6)
catalyst (15.1 & special box)
collision theory (15.5)
elementary process (15.6)
first order reaction (15.3)
initial reaction rate (15.2)
molecularity (15.6)
order of a reaction or reactant (15.3)
rate expression (15.3)
rate constant (15.3)
rate determining step (15.6)
reaction energy diagram (15.5)
reaction rate (15.2)
reaction mechanism (15.6)
reaction intermediate (15.6)
reaction half-life (15.4)
second order reaction (15.3)
termolecular (15.6)
third order reaction (15.3)
total reaction order (15.3)
transition state energy (15.5)
transition state theory (15.5)
unimolecular (15.6)
zero order reaction (15.3)

CONCEPT TEST

1. If we plot the concentration of one of the reactants in a chemical reaction against the time, the rate of the process at a given time is equal to the _____.

2. The initial reaction rate is important because at the start there are no _____ present to allow the reverse reaction to replace the reactants.

3. If for the equation

$$2 A + B \rightarrow A_2B$$

the rate expression is

$$Rate = k [A][B]$$

what is the order of the reaction? _____

4. The half-life of a certain first-order reaction is 2.2 hours. What fraction of the original reactant in this process will remain after 8.8 hours? _____

5. What is the value of the rate constant for a reaction that has a half-life of 2.2 hours?

_____

6. For a first order reaction, you will obtain a straight line if you plot _____ against

time; for a second order reaction, you must plot
_____ against time to obtain a straight line.

7. Before products are produced, reactants usually collide to form an unstable intermediate species called a transition state or a(n) _____.

8. In order to react, molecules must have an energy in excess of the _____ for the reaction.

9. According to the Arrhenius equation, we should obtain a straight line if we plot ln k against _____.

10. The collection of elementary steps that produces an overall reaction is called the _____.

11. _____ is the molecularity of the elementary process

$$2 A + B \rightarrow A_2B.$$

How probable is it that this elementary process will occur? _____.

12. When measuring the overall rate of a chemical reaction, you are usually actually measuring the rate of a single step in the mechanism called the _____.

13. A reagent that accelerates a chemical reaction but is not, itself, transformed in the process, is called a(n) _____.

14. When a change in the concentration of a reactant has no effect on the rate of the reaction, the order of that reactant in the rate expression is _____.

STUDY HINTS:

1. Although the exponents in the rate expression will only be the same as the stoichiometric coefficients in the balanced equation by coincidence, it's important to remember that the exponents in the rate expression for an elementary process are always equal to the coefficients of the reactants in that elementary process.

2. There are three basic types of problems that involve the determination of the overall rate of a chemical reaction. Although these situations are quite different, students will sometimes confuse them. In the first case, you will be given _initial_ rates and reactant concentrations from _several different experiments_; this type of problem is best solved by inspection. In the second type, you are told the concentration of a reactant at _a number of different times_; a graphical solution is usually best here. A third case involves using a proposed mechanism to predict a rate expression. When you encounter one of these problems, look carefully at what data is given, and that will usually lead you to the correct method of solution.

Don't confuse the determination of reaction order with other problem types. The most common source of confusion is the use of the Arrhenius equation to determine the variation of reaction rate with temperature. Observe carefully how these problem types differ, and it should be easy to recognize them as different cases.

3. Read carefully when doing problems that use the integrated rate equation. The information provided may be either the amount that has reacted or the amount that remains. Remember that the integrated rate equation requires the concentration at some specific time, indicated by the value of t. Unless you are cautious, it's easy to confuse the amount that has reacted with the amount that didn't react. This is a common source of errors in this problem type. Practice problems 3b and 3c are examples of this situation.

4. When using a proposed reaction mechanism to predict a rate expression, remember that this prediction is only a theory. If it disagrees with the experimental rate expression, it is necessary to revise the mechanism. On the other hand, the fact that a certain mechanism does agree with the experimental rate expression doesn't insure that the mechanism is correct. For many years the reaction

$$2 \; HI(g) \quad \rightarrow \quad H_2(g) \quad + \quad I_2(g)$$

was given as a classic example of a bimolecular reaction, but it was finally discovered that the mechanism was much more complicated (J. Chem. Phys., **36**, 1925, (1962)).

5. Remember that mathematically

$$\ln A/B = \ln A - \ln B$$

In some problems you will need to use this relationship to solve for either the initial concentration or the concentration at the given time.

6. Remember that the R value in the Arrhenius equation has different units (and value) from that which you have used with the gas laws.

PRACTICE PROBLEMS

1. (**L2**) On the left of the table on the next page are various possible changes in the concentration of a reactant, A. Fill in the blanks by determining the effect of the change indicated on a reaction that has the rate expression given at the top of the column.

Change in [A]	Rate = [A]	Rate = $[A]^2$
double [A]	_____	_____
triple [A]	_____	_____
halve [A]	_____	_____
one-fourth [A]	_____	_____
quadruple [A]	_____	_____

2. (**L2**) The data below was collected for the hydrolysis of a simple sugar in aqueous solution at 23 degrees Celsius.

[sugar]	time
2.00 mmol/dm^3	0 min
1.62	60
1.31	120
0.853	240
0.364	360

Determine whether this is a first or second order reaction by obtaining a separate sheet of graph paper and graphing (a) ln [sugar] vs. time and then (b) 1/[sugar] vs. time. In which case do you obtain a straight line? What is the order of the reaction? From the slope of the line, what is the value of k for this reaction?

3. (**L3**) a. The rate constant is 2.84×10^{-5} yr^{-1} for the first-order radioactive decay of plutonium-239. a. What is the half-life for this isotope? b. How long must a given sample of this isotope remain undisturbed until the amount of plutonium remaining is 20.0% as great as the original amount?

250

c. Based on the information provided in part a, how long must a sample of plutonium-239 remain undisturbed in order to allow 20.0% of the original plutonium to undergo radioactive decay?

4. (L2) The decomposition of gaseous hydrogen iodide at 716°C is described by the equation

$$HI_{(g)} \quad \rightarrow \quad 1/2 \ H_{2(g)} \quad + \quad 1/2 \ I_{2(g)}$$

The results of several different rate experiments are shown in the table below.

Experiment Number	Initial [HI] (mole/liter)	Initial Rate (mole/liter/min)
1	4.0×10^{-3}	1.07×10^{-9}
2	6.0×10^{-3}	2.41×10^{-9}
3	8.0×10^{-3}	4.29×10^{-9}

a. Write the rate expression for this reaction.

b. What is the overall order of this reaction?

c. Calculate the value of the rate constant for this reaction.

5. **(L6)** At 320°C the gas phase reaction

$$SO_2Cl_2 \rightarrow SO_2 + Cl_2$$

is first order with a rate constant of $2.0 \times 10^{-5} \ s^{-1}$. Which of the following mechanisms agrees with this observation?

Mechanism I: $SO_2Cl_2 \rightarrow SO_2 + Cl_2$
(simple, elementary step)

Mechanism II: $2 \ SO_2Cl_2 \rightarrow S_2O_4Cl_4$ (slow)
$S_2O_4Cl_4 \rightarrow 2 \ SO_2 + 2 \ Cl_2$ (fast)

6. **(L5)** Determine the activation energy (E_a) for the reaction

$$N_2O_{5(g)} \rightarrow 2 \ NO_{2(g)} + 1/2 \ O_{2(g)}$$

based on the following observed rate constants and temperatures:

Temperature (°C)	k (sec^{-1})
25	3.46×10^{-5}
55	1.50×10^{-3}

252

7. (**L6**) Can you suggest an explanation of why termolecular gas reactions are relatively unlikely? Under what conditions would you expect termolecular reactions to be most probable?

8. (**L2**) The elimination of methylmercury, a toxic mercury compound, from the human body is a first order process with an estimated half-life of 70 days. Individuals who eat large amounts of fish contaminated with methylmercury may have blood concentration of this compound as high as 0.200 mg/liter. If a person with this blood level of mercury stopped all mercury intake, calculate the concentration of methylmercury that would remain in the blood after 350 days.

PRACTICE PROBLEM SOLUTIONS

1.

Change in [A]	Rate = [A]	Rate = $[A]^2$
2x[A]	rate doubles	rate quadruples
3x[A]	rate triples	rate 9 times greater
1/2 [A]	rate 1/2 as great	rate 1/4 as great
1/4 [A]	rate 1/4 as great	rate 1/16 as great
4x[A]	rate 4 times as great	rate 16 times as great

2. Based on the graphs on the next page the reaction must be first order since only the ln[sugar] vs t graph gave a straight line.

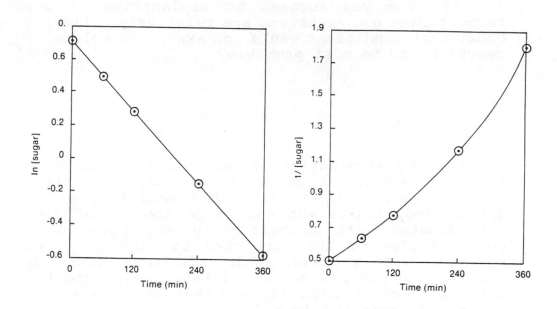

Figure 13.1 Graphs of ln[sugar] vs. t and 1/[sugar] vs. t for the hydrolysis of a sugar solution

2. (cont.) Now determine the slope of the above graph.

$$k = \frac{(\ln[\text{sugar}]_5 - \ln[\text{sugar}]_1)}{(t_5 - t_1)} = \frac{(-1.0106 - 0.69314)}{(480 - 0)}$$

$$\underline{k = 0.00355}$$

3. a.

$$t_{1/2} = \frac{0.693}{k}$$

$$t_{1/2} = \frac{0.693}{2.84 \times 10^{-5} \text{ yr}^{-1}}$$

$$\underline{t_{1/2} = 2.44 \times 10^{4} \text{yr}}$$

254

3.b. Write the integrated form of the first-order
rate equation,

$$\ln \frac{[A]}{[A_o]} = -kt$$

Notice that $[A] = 0.200 \ [A_o]$, and substitute the
values provided

$$\ln \frac{0.200 \ [A_o]}{[A_o]} = -(2.84 \times 10^{-5} \ \text{yr}^{-1})t$$

$$\underline{t = 56,700 \ \text{year}}$$

3. c. Begin again with the integrated first-order
rate equation,

$$\ln \frac{[A]}{[A_o]} = -kt$$

But this time $[A] = [A_o] - 0.200 \ [A_o] = 0.800[A_o]$.
Otherwise the values are the same.

$$\ln \frac{0.800 \ [A_o]}{[A_o]} = -(2.84 \times 10^{-5} \ \text{yr}^{-1})t$$

$$\underline{t = 7,860 \ \text{years}}$$

Notice how a small change in the wording caused a
major change in the answer.

4. a. Compare the first and third experiments. The
initial concentration of the hydrogen iodide
doubles from experiment 1 to experiment 3; however,
the rate becomes four times as great. This
indicates a second order relationship, and so the
rate equation must be

$$\text{Rate} = k[HI]^2$$

b. There is only one reactant, and so the overall
order is second order.

c. Using the data from the first experiment

$$k = Rate/[HI]^2$$

$$k = \frac{1.07 \times 10^{-9} \text{ mole/liter/min}}{(4.0 \times 10^{-3} \text{ mole/liter})^2}$$

$$\underline{k = 6.7 \times 10^{-5} \text{ (mole/liter)}^{-1}/min}$$

5. In the first mechanism, the rate is controlled by a unimolecular process, and so the reaction should be first order. This agrees with observation, and thus the first mechanism may be correct.

If the second proposed mechanism is correct, the first step controls the rate, and since that step is bimolecular, the order of that step and the overall reaction must be second order. This doesn't agree with the experimental observation, and so that mechanism cannot be correct.

6. Using the Arrhenius equation

$$\ln \frac{k_2}{k_1} = - \frac{E^*}{R} \left[\frac{1}{T_2} - \frac{1}{T_1} \right]$$

Substituting the values given

$$\ln \frac{1.50 \times 10^{-3}}{3.46 \times 10^{-5}} = - \frac{E^*}{8.314 \text{ J/mol} \cdot \text{K}} \left[\frac{1}{328K} - \frac{1}{298K} \right]$$

$$3.7694 = \frac{E^*}{8.314 \text{ J/mol} \cdot \text{K}} (3.0692 \times 10^{-4})$$

$$\underline{E_a = 1.02 \times 10^5 \text{ J/mol}} \text{ or } 102 \text{ kJ/mole}$$

7. Remember that the molecules of a gas occupy only a small fraction of the total volume of the gas sample. The large volume of empty space makes it unlikely that three molecules will simultaneously collide. In order to make these reactions more likely it is necessary to increase the number of collisions. This could be accomplished by increas-

ing the temperature, so that the molecules move faster, or by decreasing the volume, so that there is less empty space and greater probability of collisions.

8. The elimination rate constant, k_e, can be obtained from the half-life using the equation

$$k_e = \frac{0.693}{t_{1/2}} = \frac{0.693}{70 \text{ days}} = 0.0099 \text{ days}^{-1}$$

This allows you to solve the integrated first-order rate equation

$$\ln \frac{[A]}{[A_0]} = -kt$$

Substituting the values provided

$$\ln \frac{[A]}{0.200 \text{ mg/liter}} = -(0.0099 \text{ day}^{-1})(350 \text{ days})$$

$$\ln[A] = -(0.0099 \text{ day}^{-1})(350 \text{ days}) + \ln(0.200 \text{ mg/liter})$$

$$\underline{[A] = 0.006 \text{ mg/liter}}$$

PRACTICE TEST (50 Minutes)

1. What four conditions are generally considered to determine the rate of a chemical reaction?

2. At moderate conditions of temperature and pressure, each of the following reactions occurs by means of a single-step, elementary process. Write the rate expression for each of these examples, and give the order and molecularity of each.

a. $OH^- + CH_3Br \rightarrow CH_3OH + Br^-$

b. $Br(g) + H_2(g) \rightarrow HBr(g) + H(g)$

3. For the hypothetical reaction

$$A + B = C$$

The activation energy for the combination of A and B to form C is 45.0 kJ/mole. The reverse process, C forming A and B, has an activation energy of 32 kJ/mole. Is the reaction exothermic or endothermic as written above?

4. If a human ingests dichlorophenoxyacetic acid, a commonly used herbicide, the elimination in the urine can be approximated as a first-order process having a half-life of 220 hours. Suppose that a person accidentally swallows some of this compound. How much time will be required before the amount in the body is only 20.0% as great as the amount originally ingested?

5. A plot of ln K against 1/T for a certain reaction gives a straight line with a slope equal to -5300 mole K and an intercept of -24.7. Calculate the value of the activation energy, E_a, for this reaction.

6. Photochemical smog, a type of air pollution common in cities, is partially caused by reactions between ozone and hydrocarbons (emitted from automobiles). One example of this type of reaction is the combination of ozone and ethylene

$$C_2H_{4(g)} + O_{3(g)} \rightarrow H_2CO \text{ (formaldehyde)} + H_2CO_{2(g)}$$

A kinetic study of this reaction has produced the data shown in the table.

Experiment	Initial $[O_3]$	Initial $[C_2H_4]$	Initial rate
1	1.6×10^{-9} M	1.0×10^{-9} M	1.94×10^{-28} mol/L/min
2	4.0×10^{-9}	1.0×10^{-9}	4.84×10^{-28}
3	1.6×10^{-9}	3.0×10^{-9}	5.81×10^{-28}

a. Write the rate expression for this reaction.

b. What is the overall order of this reaction?

c. Calculate the initial rate of this reaction if

258

both O_3 and C_2H_4 are initially 2.0×10^{-9} M

7. For reasons best left to the imagination, certain insects, such as the common firefly, display their presence by a flashing light that is produced by a complex set of biochemical reactions. The rate at which fireflies flash varies with the temperature, since the rate-controlling step in this process is temperature dependent. It is observed that the time between flashes for a typical firefly is 16.3 seconds at 21°C, and the time between flashes is 13.0 seconds at 27.8°C. Based on this information, calculate the activation energy for the chemical reaction that controls the flashing. (Hint: Notice that you are given the time between flashes, which is inversely related to the rate of flashing.)

8. The rate expression for the reaction

$$2 \ NO(g) \ + \ 2 \ H_2(g) \ \rightarrow \ N_2(g) \ + \ 2 \ H_2O(g)$$

has the form

$$Rate = k[NO]^2[H_2]$$

Consider the following two mechanisms that have been suggested for this process:

MECHANISM I
$$NO \quad + \quad H_2 \quad \rightarrow \quad H_2O \ + \ N \quad (slow)$$

$$N \ + \ H_2 \ + \ NO \ \rightarrow \ N_2 \quad + \ H_2O \ (fast)$$

MECHANISM II
$$2 \ NO \ + \ H_2 \ \rightarrow \quad N_2 \ + \ H_2O_2 \quad (slow)$$

$$H_2O_2 \ + \ H_2 \ \rightarrow \ 2 \ H_2O \qquad \quad (fast)$$

Based on the available information, which of these mechanisms is possible?

9. At 1000°C, cyclopropane, an organic compound, reacts according to the equation

CH$_2$ H
/ \ |

$$H_2C - CH_2 \quad \rightarrow \quad H_2C = C - CH_3$$

This is a first order reaction with a half-life of 7.5×10^{-2} seconds. Calculate the length of time necessary for 90.0% of some original amount of cyclopropane to react under these conditions.

CONCEPT TEST ANSWERS

1. slope of a line tangent to the curve at that time
2. products
3. second
4. one-sixteenth
5. 0.32 hr^{-1}
6. ln [A], 1/[A]
7. activated complex
8. activation energy
9. 1/T
10. reaction mechanism
11. termolecular, not very
12. rate determining step
13. catalyst
14. zero

PRACTICE TEST ANSWERS

1. **(L1)** concentration of reactants
   temperature
   presence of a catalyst
   physical state of the reactants
2. **(L6)**
a. Rate = $- k[OH^-][CH_3Br]$, bimolecular, second order

b. Rate = $- k[Br][H_2]$, bimolecular, second order

3. **(L6)** The reaction is endothermic, with an enthalpy of reaction of 13 kJ/mole
4. **(L3)** 510 hours
5. **(L5)** 44 kJ
6. **(L2)** a. Rate = $k[O_3][C_2H_4]$

   b. Second order

   c. 4.8×10^{-28} mol/L/min
7. **(L5)** 24.5 kJ
8. **(L7)** In mechanism I, the first step is the rate controlling reaction, and based on this equation, the rate expression should be first

order in NO and first order in H_2. This is
contrary to observation, and so must be wrong.

In mechanism II, the first step is also the rate
controlling reaction, and this equation produces
a rate expression that would be second order in
NO and first order in H_2. This agrees with
observation, and may be correct.
9. (L3) 0.25 sec

CHAPTER 16
CHEMICAL EQUILIBRIUM:
GENERAL CONCEPTS

LEARNING GOALS:

1. The ability to write correct equilibrium expressions for both homogeneous and heterogeneous equilibria is essential to the material in the rest of this chapter. When given the equilibrium constant for one balanced equation, know how to obtain the new equilibrium constant if the original equation is reversed, multiplied by a constant, or combined with other balanced equations. (Sec. 16.2)
2. Equilibrium constants for gas reactions may be stated in terms of either concentrations or pressures. Be able to convert either from K_c to K_p or from K_p to K_c and to obtain the appropriate form of the equilibrium constant for a given situation. (Sec. 16.2)
3. It's important to understand the reaction quotient, since it can be a valuable tool for determining whether or not equilibrium conditions exist, and in which direction a reaction will shift to attain equilibrium. (Sec. 16.3)
4. The value of K, the equilibrium constant, can be calculated from experimental data on the percent of reaction, the concentration of the reactant that actually undergoes the process, or the equilibrium concentrations of each species in the system. Know how to calculate K in each of these cases. (Sec. 16.4)

5. If you are given the value of the equilibrium constant, there are several different types of problems that require you to calculate relationships between reactant and product concentrations. Be familiar with these various problem types. (Sec. 16.4)

6. Use le Chatelier's principle to predict how a given equilibrium system will be affected by

temperature changes, volume changes, changes in concentrations of the system components, or the addition of a catalyst. (Sec. 16.5)

7. Where appropriate, be able to use the relationship between reaction mechanism and equilibrium to obtain the rate expression from a mechanism having a fast equilibrium as the first step. (Sec. 16.5)

IMPORTANT NEW TERMS:

equilibrium constant expression (16.2)
reaction quotient (16.3)
reversible reaction (16.1)

CONCEPT TEST

1. Since the value of K depends on the stoichiometric coefficients, you must specify the _____ when using a particular K value.

2. For all reversible processes there will always be conditions of pressure and temperature for which reactants and products can _____.

3. The concentrations of what types of substances never appear in the equilibrium constant expression?

a. _____

b. _____

c. _____

4. When all of the stoichiometric coefficients in a balanced chemical equation are multiplied by a

263

constant factor, the equilibrium constant for the old equation must be _____ to obtain the new equilibrium constant.

5. When a balanced chemical equation is written in reverse, that is, the old reactants become the new products and the old products become the new reactants, the equilibrium constant for the old equation is the _____ of the new equilibrium constant.

6. When two or more balanced chemical equations are added to produce a new chemical equation, the equilibrium constant for the new equation will be equal to the _____ of the equilibrium constants for the reactions that were added.

7. A system is at equilibrium only when the reaction quotient, Q, is equal to _____.

8. The reaction quotient and the equilibrium constant expression for a given balanced equation look the same, but the difference between them is that _____.

9. What effect does a catalyst have on a system in chemical equilibrium?

_____

10. Le Chatelier's principle states that a change in a system in chemical equilibrium will cause the system to change in such a manner as to _____ _____ the effect of the change.

STUDY HINTS:

1. When students first begin to write equilibrium expressions, it is sometimes tempting to include the plus sign from the balanced equation, which is, of course, wrong. Thus for the equation

$$A + B \rightleftharpoons C$$

They might write an equilibrium expression in the <u>incorrect</u> form

$$K = \frac{[C]}{[A] + [B]} \qquad \text{INCORRECT!}$$

The mistake comes from writing down the reactants without thinking. More experience (and loss of points on some tests) quickly cures this error, but it's better to be sure to avoid this mistake from the beginning.

2. Most students quickly recognize how to use Le Chatelier's principle to predict the effects of changing system conditions on a chemical equilibrium, but this type of question frequently contains tricks that will trap the unwary. Remember that pure solids, pure liquids, and solvents in dilute solutions don't appear in the equilibrium expression. This also means that adding more of these substances to (or removing them from) an equilibrium system will have no effect on the equilibrium.

Also notice that the shift in the equilibrium doesn't completely eliminate the original stress. For example, if I_2 is added to the system

$$2 HI(g) \rightleftharpoons H_2(g) + I_2(g)$$

the equilibrium does shift to the left, and the concentrations of hydrogen and iodine gases are

265

less than they were <u>after the extra iodine was added</u>. It is important to realize that the final concentration of iodine gas is greater than it was <u>before</u> you added the extra iodine.

3. As the name implies, an equilibrium expression is not a true equality unless all of the concentration values used are equilibrium concentrations. As the problems become more complicated, keep in mind that the ultimate purpose of what we are doing is usually to obtain equilibrium values so that we can use them in the equilibrium expression.

4. Only concentrations or pressures can be used in the equilibrium expression. In some problems the number of moles and the volume of the container or the number of moles and the volume of the solution are given rather than concentrations. Don't forget to convert these values to concentration units. Whenever a problem statement provides the volume of the aqueous solution or the volume of a gas container, check to see if a conversion is necessary to obtain molar concentrations.

5. This is a good time to review the method for solving a quadratic equation, since this skill will be necessary for some of the problems in this chapter.

PRACTICE PROBLEMS

1. (L6) The following mixture is at equilibrium in a closed container

$$Ti(s) + 2 Cl_2(g) \rightleftharpoons TiCl_4(g) + energy$$

Listed on the left below are several changes that may affect the equilibrium composition of this system. For each separate change indicated, place an X in the appropriate space on the right to indicate whether that single change will increase, decrease, or have no effect on the concentration of chlorine gas present in the system after equilibrium has been restored.

	increase [Cl$_2$]	decrease [Cl$_2$]	no effect
a. Add TiCl$_4$ to the system	_____	_____	_____
b. Add Ti to the system	_____	_____	_____
c. Decrease the temperature	_____	_____	_____
d. Add an inert gas to increase the pressure	_____	_____	_____
e. Add Cl$_2$ to the system	_____	_____	_____

2. **(L1)** Write equilibrium expressions for each of the following reactions:

a. $Br_2(g) \rightleftharpoons 2\ Br(g)$

b. $H_2(g) + I_2(g) \rightleftharpoons 2\ HI(g)$

c. $2\ NOBr(g) \rightleftharpoons 2\ NO(g) + Br_2(g)$

d. $TiCl_4(g) \rightleftharpoons Ti(s) + 2\ Cl_2(g)$

e. $2\ HgO(s) \rightleftharpoons 2\ Hg(s) + O_2(g)$

f. $Ag^+(aq) + 2\ NH_3(aq) \rightleftharpoons Ag(NH_3)_2^+(aq)$

3. (**L2**) The value of K_p is 1.2 atm^{-2} at 377°C for the reaction

$$3 H_2(g) + N_2(g) \rightleftharpoons 2 NH_3(g)$$

Calculate the value of K_c for this reaction.

4. (**L4**) Calculate the value of the equilibrium constant, K_c, for the reaction

$$H_2(g) + Br_2(g) \rightleftharpoons 2 HBr(g)$$

if the following equilibrium concentrations are observed at 1297 K, $[H_2] = [Br_2] = 0.020$ M and $[HBr] = 7.4$ M

5. (**L6**) Silver chloride and silver iodide are both relatively insoluble in water. The equilibrium reactions and constants for the dissolving of these compounds are as follows:

$$AgCl(s) \rightleftharpoons Ag^+(aq) + Cl^-(aq) \qquad K = 1.8 \times 10^{-10}$$

$$AgI(s) \rightleftharpoons Ag^+(aq) + I^-(aq) \qquad K = 1.5 \times 10^{-16}$$

If solid AgCl and solid AgI are placed in water in separate beakers, which beaker will have the greater concentration of Ag^+?

6. (**L4**) When 9.500 moles of PCl_5 gas is placed in a 2.00 liter container and allowed to come to equilibrium at a constant temperature, it is observed that 40.0% of the PCl_5 decomposes according to the equation

$$PCl_5(g) \quad \rightleftharpoons \quad PCl_3(g) \quad + \quad Cl_2(g)$$

What is the value of K_c for this reaction under these conditions?

7. (**L5**) When 1.00 mole of phosphorus pentachloride gas is placed in a 2.00 liter container at $250°C$ and allowed to react according to the equation

$$PCl_5(g) \quad \rightleftharpoons \quad PCl_3(g) \quad + \quad Cl_2(g)$$

the equilibrium constant for the process, K_c, has a value of 0.0415 M. Calculate the equilibrium concentration for each species in this system.

8. (L5) Calculate the equilibrium concentrations that would result in the previous problem if you initially added not only 1.00 mole of phosphorus pentachloride but also 0.200 moles of chlorine gas to the original system and allowed enough time to pass so that equilibrium conditions were restored.

9. (L5) At 400°C, the reaction below occurs when 1.50 moles of $POCl_3$ is placed in a 0.500 liter container

$$POCl_3(g) \rightleftharpoons POCl(g) + Cl_2(g) \qquad K_c = 0.248 \text{ M}$$

Calculate the number of moles of POCl that must be added to this system in order to produce an equilibrium concentration of Cl_2 equal to 0.500 M.

10. **(L3)** At 1000 K, $3.1 \times 10^{-3} = K_p$ for the reaction

$$I_2(g) \rightleftharpoons 2 I(g)$$

It is observed that in a certain sealed container at $25^{\circ}C$ the pressure of I_2 gas is 0.21 atmospheres and the pressure of I gas is 0.030 atmospheres. a. Is this system at equilibrium? If it is not at equilibrium, will the pressure of I_2 increase or decrease as it continues to approach equilibrium? b. If a catalyst is added to the system, what effect will it have on the equilibrium?

11. **(L7)** For the reaction

$$H_2(g) + I_2(g) \rightleftharpoons 2 HI(g)$$

The rate expression is reaction rate = $k[H_2][I_2]$

Does the following proposed mechanism agree with the rate expression?

$$I_2 \underset{k_{-1}}{\overset{k_1}{\rightleftharpoons}} 2 I \qquad \text{fast, equilibrium}$$

$$2 I + H_2 \overset{k_2}{\rightarrow} 2 HI \qquad \text{slow, rate determining}$$

PRACTICE PROBLEM SOLUTIONS

1.

	increase $[Cl_2]$	decrease $[Cl_2]$	no effect
a. Add $TiCl_4$ to the system	X		
b. Add Ti to the system			X
c. Decrease the temperature		X	
d. Add an inert gas to increase the pressure		X	
e. Add Cl_2 to the system	X		

2. a. $K = \dfrac{[Br]^2}{[Br_2]}$ b. $K = \dfrac{[HI]^2}{[H_2][I_2]}$

c. $K = \dfrac{[NO]^2[Br_2]}{[NOBr]^2}$ d. $K = \dfrac{[Cl_2]^2}{[TiCl_4]}$

e. $K = [O_2]$ f. $K = \dfrac{[Ag(NH_3)_2^{+}]}{[Ag^{+}][NH_3]^2}$

3. The relationship between K_p and K_c is

$$K_p = K_c(RT)^{\Delta n}$$

Δn = moles of gas produced - moles of gas reacted
      = 2 - 4 = - 2

$K_c = (1.2 \text{ atm}^{-2})(0.08206 \text{ L·atm/mole·K} \times 650 \text{ K})^2$

$\underline{K_c = 3.4 \times 10^3}$

4. A data table is useful here, even though the only values given are equilibrium concentrations, and so the data table has only one line.

	$[H_2]$	$[Br_2]$	$[HBr]$
equilibrium concentrations	0.020	0.020	7.4

Write the equilibrium expression

$$K_c = \frac{[HBr]^2}{[H_2][Br_2]}$$

Substitute the values from the equilibrium line of the data table

$$K_c = \frac{(7.4)^2}{(0.020)(0.020)}$$

$$\underline{K_c = 1.4 \times 10^5}$$

5. The product of the ion concentrations is much smaller for AgI than for AgCl. Since both compounds produce the same number of moles of ions per mole in solution, this suggests that the individual ion values will be larger for AgCl, and so the silver ion concentration will be larger in the beaker containing the silver chloride.

6. The first thing that should catch your attention is the fact that the volume of the container is provided. With this clue, it is easy to see that you are given moles of gas, not concentration of gas. Begin by calculating the initial concentration of PCl_5.

$$[PCl_5] = 9.500 \text{ moles}/2.00 \text{ liters} = 4.750 \text{ M}$$

If 40.0% of this gas reacts,
the concentration reacting = 0.400 x 4.750 M
= 1.896 M

You are now ready to write the data table for the reaction.

Data Table

	$[PCl_5]$	$[PCl_3]$	$[Cl_2]$
Before dissociation	4.750	0	0
Change	-1.896	+1.896	+1.896
Equilibrium conc.	2.854	1.896	1.896

Write the equilibrium expression

$$K_c = \frac{[PCl_3][Cl_2]}{[PCl_5]}$$

Substitute the values from the equilibrium concentration line of the data table

$$K_c = \frac{(1.896)(1.896)}{(2.854)}$$

$$\underline{K_c = 1.26}$$

7. a. Begin by calculating the initial concentration of the PCl_5.

initial $[PCl_5]$ = 1.00 mole/2.00 liter = 0.500 M

Now set up the data table, letting X = $[PCl_5]$ that reacts

	$[PCl_5]$	$[PCl_3]$	$[Cl_2]$
Initial conc.	0.500	0	0
Change	-X	+X	+X
Equilibrium conc.	0.500-X	X	X

Now write the equilibrium expression

$$K_c = \frac{[PCl_3][Cl_2]}{[PCl_5]}$$

Substitute the equilibrium concentrations from the data table.

$$0.0415\ M = \frac{(X)(X)}{(0.500-X)}$$

Cross multiply and rearrange

$$X^2 + 0.0415X - 0.02075 = 0$$

Compare this with the general quadratic form

$$aX^2 + bX + c = 0$$

and substitute the values for a, b, and c into the equation

$$X = \frac{-b + \sqrt{b^2 - 4ac}}{2a}$$

$$X = \frac{-0.0415 + \sqrt{(0.0415)^2 - 4(-0.02075)}}{2}$$

$$X = 0.125$$

Thus from the data table,

$$[PCl_5] = 0.500 - 0.125 = \underline{0.375 \text{ M}}$$

$$[PCl_3] = [Cl_2] = \underline{0.125 \text{ M}}$$

8. The initial concentration of the PCl_5 is still the same, and

initial $[Cl_2] = 0.200$ mole/2.00 liter $= 0.100$ M

Now set up the data table, letting X = $[PCl_5]$ that reacts

	$[PCl_5]$	$[PCl_3]$	$[Cl_2]$
Initial conc.	0.500	0	0.100
change	-X	+X	+X
Equilibrium conc.	0.500-X	X	0.100+X

Now write the equilibrium expression

$$K_c = \frac{[PCl_3][Cl_2]}{[PCl_5]}$$

Substitute the equilibrium concentrations from the data table.

$$0.0415 \text{ M} = \frac{(X)(0.100+X)}{(0.500-X)}$$

Cross multiply and rearrange

$$X^2 + 0.1415X - 0.02075 = 0$$

Compare this with the general quadratic form

$$aX^2 + bX + c = 0$$

and substitute in the quadratic equation as in the previous problem

$$X = \frac{-0.1415 \pm \sqrt{(0.1415)^2 - 4(-0.02075)}}{2}$$

$$X = 0.0897$$

$[PCl_5] = 0.500 - 0.0897 = \underline{0.410 \text{ M}}$

$[PCl_3]$ $\underline{0.0897 \text{ M}}$

$[Cl_2] = 0.100 + 0.0897 = \underline{0.1897 \text{ M}}$

Notice that the effect has been to shift the equilibrium to the left, exactly as Le Chatelier's principle would predict.

9. Calculate the initial concentration of $POCl_3$

$$[POCl_3] = 1.50 \text{ moles}/0.500 \text{ liters} = 3.00 \text{ M}$$

Now write the data table, letting X = [POCl] added

	$[POCl_3]$	$[POCl]$	$[Cl_2]$
Original conc.	3.00	X	0
Change	-0.500	+0.500	+0.500
Equilibrium conc.	3.00-0.500 =2.50	X+0.500	0.500

276

Notice that because the equilibrium concentration of chlorine was known to be 0.500 M, this required the reaction of 0.500 M of the $POCl_3$ and formation of 0.500 M of the POCl.

Now write the equilibrium expression

$$K_c = \frac{[POCl][Cl_2]}{[POCl_3]}$$

Substitute the equilibrium concentrations from the data table

$$0.248 = \frac{(X+0.500)(0.500)}{(2.50)}$$

Solving

$$0.620 = 0.500X + 0.25$$

$$X = 0.740 \text{ M POCl}$$

The problem is not completed, however, since this is the equilibrium concentration of POCl and you were asked to find the amount added.

moles of POCl added = 0.740 M x 0.500 liters

<u>moles of POCl added =0.370 moles</u>

10. First, as usual, write the data table, but this time it's based on pressures.

	$P(I_2)$	$P(I)$
Observed pressure	0.21 atm	0.030 atm

Insert these values into the reaction quotient

$$Q = \frac{P_I^2}{P_{I_2}}$$

$$Q = \frac{(0.030 \text{ atm})^2}{(0.21 \text{ atm})}$$

$$Q = 4.3 \times 10^{-3}$$

a. Since $Q > K_p$, <u>the system is not in equilibrium.</u>
     As equilibrium is approached, since $Q > K$, the equilibrium will shift to the left, and the pressure of I_2 will increase.

b. <u>The system may attain equilibrium more rapidly, but the position of the equilibrium will not change.</u>

11. Based on the slow step, the rate expression is

$$\text{rate} = k_2[I]^2[H_2]$$

To compare this rate expression with the one given, it must be converted so that it's in terms of the original reactants. The proposed mechanism indicates that I is formed rapidly from I_2, and the rate of that reaction is

$$\text{rate of production of I} = k_1[I_2]$$

and since I is removed very slowly, most of it will be converted back to I_2 before it can react with the H_2. This rate is

$$\text{rate of reversion of I to } I_2 = k_{-1}[I]^2$$

At equilibrium, the forward and reverse rates must be equal, and so

$$k_1[I_2] = k_{-1}[I]^2$$

Rearranging this expression

$$[I]^2 = \frac{k_1[I_2]}{k_{-1}}$$

Now substitute this value for $[I]^2$ into the rate expression obtained for the slow step,

$$\text{rate} = k_2[I]^2[H_2] = k_2[H_2] \times \frac{k_1[I_2]}{k_{-1}}$$

Combining all of the constants to produce one value

$$\text{rate} = k[H_2][I_2]$$

Since this is the experimentally obtained rate law, this mechanism is reasonable.

PRACTICE TEST (55 min.)

1. Consider the following two equilibrium reactions

Reaction 1 $2 \text{ HI(g)} \rightleftharpoons H_2(g) + I_2(g)$

Reaction 2 $1/2 \ H_2(g) + 1/2 \ I_2(g) \rightleftharpoons \text{HI(g)}$

If the equilibrium constant for reaction 1 is K_1, and the equilibrium constant for reaction 2 is K_2, what is the mathematical relationship between K_1 and K_2?

a. $K_1 = K_2^2$ b. $K_1 = 1/K_2$

c. $K_1 = -K_2$ d. $K_1 = 1/K_2^2$

2. At 275°C, $K_p = 1.14 \times 10^3$ atm^2 for the equilibrium system

$$CH_3OH(g) \rightleftharpoons CO(g) + 2 \ H_2(g)$$

What is the value of K_c for this reaction under these conditions?

3. At 400°C, $K_c = 7.0$ for the equilibrium

$$Br_2(g) + Cl_2(g) \rightleftharpoons 2 \ BrCl(g)$$

If 0.060 moles of bromine gas and 0.060 moles of chlorine gas are introduced into a 1.00 liter container at 400 degrees Celsius, what is the concentration of BrCl when sufficient time has passed so that equilibrium is established?

279

4. The following mixture is at equilibrium in a closed container

$$NH_4HS(s) \rightleftharpoons NH_3(g) + H_2S(g) + energy$$

Listed on the left below are several changes that may affect the equilibrium composition of this system. For each separate change indicated, place an X in the appropriate space on the right to indicate whether that single change will increase, decrease, or have no effect on the concentration of hydrogen sulfide gas present in the system after equilibrium has been restored.

	increase [H_2S]	decrease [H_2S]	no effect
a. Add NH_3 to the system	_____	_____	_____
b. Add NH_4HS to the system	_____	_____	_____
c. Increase the temperature	_____	_____	_____
d. Add an inert gas to increase the pressure	_____	_____	_____
e. Add H_2S to the system	_____	_____	_____

5. At 2727°C, K_c = 0.37 for the reaction

$$Cl_2(g) \rightleftharpoons 2 Cl(g)$$

At equilibrium in a closed container, the pressure of Cl_2 = 0.86 atmospheres. What is the pressure of $Cl(g)$ in this container?

6. Although coal was initially the major source of raw materials for the chemical industry, it became

less important as petroleum and natural gas became readily available. As the oil supply decreases, coal is expected to again become important. The gasification of coal will probably play an important role, and a reaction called the water-gas shift reaction is an important step in our current efforts to convert coal into gaseous products.

The water-gas shift reaction is shown below, and in one version of this process the equilibrium constant for this reaction has a value of 2.23.

$$CO(g) + H_2O(g) \rightleftarrows CO_2(g) + H_2(g)$$

The table below lists the concentrations of these four reactants in a number of experiments. In each case, compare the value of the reaction quotient with the value of the equilibrium constant (2.23) to determine if the system is at equilibrium, and if it is not at equilibrium, to determine which way the reaction will probably proceed.

	Concentrations				Prediction
	$[CO]$	$[H_2O]$	$[CO_2]$	$[H_2]$	
a.	0.951	0.432	1.38	0.773	_____
b.	0.389	0.182	0.510	0.309	_____
c.	0.749	0.356	0.960	0.567	_____

7. If 0.800 moles of HI gas is placed in a 2.00 liter container and allowed to come to equilibrium at a constant temperature, it will dissociate according to the equation

$$2 HI(g) \rightleftarrows H_2(g) + I_2(g)$$

When equilibrium is attained, the concentration of iodine gas in the container is 0.100 Molar. What is the value of K_c for this reaction under these conditions?

8. Hydrogen sulfide gas is placed in a 2.00 liter

container at an elevated temperature and reacts according to the equation

$$2 H_2S(g) \rightleftharpoons 2 H_2(g) + S_2(g)$$

Calculate the value of K_c for this reaction if the resulting gas mixture consists of 1.0 moles of H_2S, 0.20 moles of H_2, and 0.80 moles of S_2.

CONCEPT TEST ANSWERS

1. balanced equation
2. coexist at equilibrium
3. a. pure solids   b. pure liquids   c. solvents in dilute solutions   (These may be in any order.)
4. raised to a power equal to the multiplying factor
5. reciprocal   6. product
7. the equilibrium constant
8. the concentrations in the reaction quotient expression are not necessarily equilibrium concentrations.
9. It changes the rate of approach to equilibrium conditions, but does not affect the composition of the equilibrium mixture.
10. reduce or counteract

PRACTICE TEST ANSWERS

1. **(L1)** d. $K_1 = 1/K_2^2$

2. **(L2)** $K_c = 0.565$ M^2

3. **(L5)** [BrCl] = 0.044 M

4. **(L6)**

	increase [H_2S]	decrease [H_2S]	no effect
a. Add NH_3 to the system		X	
b. Add NH_4HS to the system			X
c. Increase the temperature		X	

d. Add an inert
   gas to increase
   the pressure _____ ___X___ _____
e. Add H_2S to
   the system ___X___ _____ _____

5. (**L5**) $P_{Cl} = 8.9$ atm
6. (**L3**)
a. Q = 2.61, since Q > K the reaction will shift to
   the left.
b. Q = 2.23, since Q = K the reaction is at
   equilibrium.
c. Q = 2.04, since Q < K the reaction will shift to
   the right.
7. (**L4**) $K_c = 0.25$
8. (**L4**) $K_c = 0.016$

CHAPTER 17
THE CHEMISTRY OF
ACIDS AND BASES

LEARNING GOALS:

1. Be familiar with the auto-ionization of water and the role it plays in our understanding of acids and bases. Know the Bronsted definitions of acids and bases, be able to identify the conjugate acid-base pairs in an acid-base reaction, and be able to determine the relative strength of acids or bases using Table 17.2. (Secs. 17.1 & 17.2)

2. The equilibrium expression for the auto-ionization of water provides an excellent method for relating the hydrogen ion and hydroxide ion concentrations to each other.

$$K_w = [H_3O^+][OH^-]$$

This will be useful in many problems. Understand also how the relationship between acidity and basicity can also be represented using the pH scale. (Sec. 17.3)

3. The list of strong acids and bases is relatively short. Memorizing this list of compounds will make it easier to recognize the problem types in this chapter. (Sec. 17.4)

4. Many of the problems in this chapter are based on weak acid and weak base equilibria. Degree of ionization may be indicated by the percent dissociation, ionization constant, or pH. If the initial concentration of a weak acid or base and any one of these three quantities is given, be able to calculate the missing values. This may require a review the methods for solving quadratic equations. (Sec. 17.5)

5. Be familiar with the special types of acid-base equilibria described in this chapter including

ionization of polyprotic acids and ionization of
salts. Although these problems use the same basic
principles previously learned, they do present some
minor differences that can be important. (Sec.
17.5)

6. Understand the Lewis theory of acids and bases,
recognize how it differs from the Bronsted theory,
and use it to identify acids, bases and adducts in
chemical reactions. It's also necessary to be able
to use formation constants to measure the extent of
these reactions. (Sec. 17.6)

IMPORTANT NEW TERMS:

adduct or complex (17.6)
amphiprotic (17.2)
amphoteric (17.6)
auto-ionization (17.1)
Bronsted acid or base (17.2)
conjugate acid or base (17.2)
conjugate acid-base pair (17.2)
coordinate covalent bond (17.6)
coordination number (17.6)
formation constant (17.6)
hydrolysis reaction (17.5)
leveling effect (17.4)
Lewis acid or base (17.6)
monoprotic acid (17.2)
neutral solution (17.3)
pH scale (17.3)
polyprotic acid (17.5)
strong acid or base (17.4)
weak acid or base (17.5)

CONCEPT TEST

1. Arrhenius proposed that a substance producing

_____ as one of the products of ionic

dissociation in water should be called an acid, and

a substance producing _____ should be

called a base.

2. The Bronsted definition of acids and bases states that an acid is a(n) _____, and a base is a(n) _____.

3. Acids capable of donating more than one proton are called _____ acids; substances that can act as either a Bronsted acid or base are called _____.

4. Identify the conjugate acid-base pairs.

$$H_2PO_4^- \ + \ H_2O \ \rightleftarrows \ H_3O^+ \ + \ HPO_4^{2-}$$

5. Bronsted acids that ionize 100% in water are called _____ acids; bases that ionize 100% in water are called _____ bases.

6. Label each compound as a strong acid, strong base, weak acid, or weak base.

a. HNO_3 _____   b. HOAc _____

c. NaOH _____   d. HCl _____

7. Hydrofluoric acid, HF, is a weak acid. Based on this information, which of the following statements is most likely to be correct?

a. HF won't hurt if it's spilled on the skin.

b. F^- is a strong conjugate base.

c. HF is 100% ionized in aqueous solution.

8. At 25°C, pH + pOH must equal _____ for aqueous solutions.

9. _____ is the strongest acid that can exist in water. What is the strongest base that can exist in water? _____

10. For polyprotic acids, the pH of the solution depends primarily on the hydronium ion generated in the _____.

11. In general, increasing the number of oxygen atoms in an oxyacid _____ (choose from increases or decreases) the relative acid strength.

12. The Lewis definition states that an acid is a(n) _____, and a base is a(n) _____.

13. Notice that the defining equation for formation constants is exactly the reverse of the equation used to define the corresponding dissociation or instability constant. This means that the value of the formation constant will be equal to the _____ of the value for the dissociation constant for the same compound.

14. A substance that is a Bronsted base and Lewis acid is said to be _____ . (Contrast

this with the definition of amphiprotic.)

STUDY HINTS:

1. Notice that the pH scale is not limited to values between 1 and 14. Although most of the solutions commonly encountered will fall into that range, values higher than fourteen and lower than one are possible. To prove this, calculate the pH of a 1 M solution of KOH.

2. Remember that the pH of a neutral solution is 7 only when the solution temperature is 25 degrees Celsius (see practice test problem 6). In a neutral solution it is always true that hydronium ion concentration equals hydroxide ion concentration, and so this is a better definition of a neutrality in aqueous solutions.

3. As was the case in the previous chapter, a data table is an extremely valuable way to organize the information in these equilibrium problems. If it isn't already a habit to use a data table, form that habit now.

4. For many of the problems in this chapter, the first step in the solution is to identify the type of compound involved. Strong acids and bases are 100% ionized; there is no equilibrium expression. Weak acids and bases ionize slightly, so an equilibrium equation and expression is essential. In order to determine the pH of salt solutions, it's necessary to identify the ion that undergoes hydrolysis and then write the equilibrium expression and equation for that species. Failure to determine what type of substance is involved can make this problem type difficult to identify.

5. Is it possible to make a basic solution by adding a weak acid to pure water? Of course not, but in some cases the calculations will seem to indicate that the pH of an extremely weak base is less than 7 or of an extremely weak acid is more than 7. In these cases, water becomes the species that controls the pH, and the solution is essentially neutral. Always remember to check and make sure that the answer is reasonable.

6. Problems that involve formation constants often involve taking a third, fourth, or higher root. If this procedure is not already familiar, be sure to look it up in the instruction book for the calculator or ask the instructor for help.

PRACTICE PROBLEMS

1. **(L1)** For each reaction, identify the conjugate acid-base pairs and then use Table 17.2 from the textbook to predict whether the equilibrium lies predominantly to the left or to the right.

a. $H_2PO_4^-(aq) + H_2S(aq) \rightleftharpoons H_3PO_4(aq) + HS^-(aq)$

b. $HCO_3^-(aq) + NH_4^+(aq) \rightleftharpoons NH_3(aq) + H_2CO_3(aq)$

2. **(L1)** a. What is the conjugate base
of HPO_4^{2-}? _____

b. What is the conjugate acid
of HPO_4^{2-}? _____

3. **(L1)** For plants that require acidic soil, special fertilizers are sold that include metal ion compounds to increase soil acidity. Examine Table 17.6 in the text and predict which of the metal ions listed there will be most effective in lowering the soil pH.

4. **(L2)** Heavy water is formed by replacing normal hydrogen atoms with a heavier hydrogen isotope, deuterium (symbol D), that has an atomic mass of 2.0. Chemically it is very similar to normal water, and the equation for the auto-ionization of heavy water, D_2O, may be written

$$D_2O \rightleftharpoons D^+ + OD^-$$

At 25°C, $K_w = 1.11 \times 10^{-15}$ M^2 for this reaction. What is the pD of a neutral solution of heavy water at 25 degrees Celsius? Assume that pD is similar to pH, that is, pD = - log [D$^+$].

5. **(L1)** Supply the missing values in the following table (at 25°C):

pH	pOH	[H$^+$]	[OH$^-$]
2.45			
	5.42		
			9.2×10^{-3}
		6.5×10^{-5}	

6. **(L1)** Assuming equal concentrations of each substance in aqueous solution, arrange the following in order of increasing acidity: $NaC_2H_3O_2$, NH_4Cl, and $HC_2H_3O_2$.

least
(acidic) _____ < _____ < _____ (acidic)   most
(acidic)

7. **(L3)** Calculate the pH and pOH of a 0.50 M solution of HNO_3.

8. (**L4**) The ionization equation for hydrazoic acid HN_3, is

$$HN_3 \rightleftharpoons H^+ + N_3^-$$

At $25°C$ $[H^+] = 1.38 \times 10^{-3}$ M for a 0.10 M aqueous solution of hydrazoic acid. Calculate the K_a for this weak acid under these conditions.

9. (**L4**) If $K_b = 7.4 \times 10^{-5}$ for trimethylamine, a weak base that ionizes as shown in the equation below, calculate the pH of a 0.20 M solution of this substance.

$$(CH_3)_3N(aq) + H_2O(\ell) \rightleftharpoons (CH_3)_3NH^+(aq) + OH^-(aq)$$

10. (**L6**) For each of the following compounds, predict whether it is most likely to act as a Lewis acid or a Lewis base.

a. H_2Se _____ b. AsH_3 _____

c. Ag^+ _____ d. Ti^{4+} _____

11. **(L4)** Calculate the pH of a 0.05 M solution of hydrofluoric acid, HF, given the following dissociation equation:

$$HF(aq) \rightleftharpoons H^+(aq) + F^-(aq) \qquad K_a = 7.2 \times 10^{-4}$$

12. **(L6)** The formation constant is 1.3×10^7 for the reaction

$$Ag^+ + 2\ Br^- \rightleftharpoons [AgBr_2^-]$$

When 25.0 mL of 0.010 M Ag^+ is mixed with 25.0 mL of 0.20 NaBr, what is the equilibrium concentration of silver ion that results?

PRACTICE PROBLEM SOLUTIONS

1. a. $H_2PO_4^-(aq) + H_2S(aq) \rightleftharpoons H_3PO_4(aq) + HS^-(aq)$

    base$_1$ acid$_2$ acid$_1$ base$_2$

Since H_3PO_4 is a stronger proton donor than H_2S, this equilibrium will go to the left.

b. $HCO_3^-(aq) + NH_4^+(aq) \rightleftharpoons NH_3(aq) + H_2CO_3(aq)$

    base$_1$ acid$_2$ base$_2$ acid$_1$

Since H_2CO_3 is a stronger proton donor than NH_4^+, this equilibrium will also go to the left.

2. a. PO_4^{3-} is the conjugate base of HPO_4^{2-}

   b. $H_2PO_4^-$ is the conjugate acid of HPO_4^{2-}

3. Al^{3+} and Fe^{3+} have the largest hydrolysis constants and are commonly used to increase soil acidity for plants.

4. Write the equilibrium expression for the equilibrium reaction given.
$$K_w = [D^+][OD^-]$$

In order for the solution to be neutral,
$$[D^+] \text{ must equal } [OD^-]$$

and so the equilibrium expression becomes
$$K_w = [D^+]^2$$
Inserting the available values

$$1.11 \times 10^{-15} = [D^+]^2$$
or
$$[D^+] = 3.33 \times 10^{-8}$$
Substitute this value into the pD expression given

$$pD = -\log[D^+] = -\log(3.33 \times 10^{-8})$$

$$\underline{pD = 7.48}$$

5.

pH	pOH	[H$^+$]	[OH$^-$]
2.45	11.55	3.5x10^{-3}	2.8x10^{-12}
8.58	**5.42**	2.6x10^{-9}	3.8x10^{-6}
11.96	2.04	1.1x10^{-12}	**9.2x10^{-3}**
4.19	9.81	**6.5x10^{-5}**	1.5x10^{-10}

6. First, identify the acidic species supplied by each compound. $HC_2H_3O_2$ is itself a weak acid, NH_4^+ and Na^+ are the acidic conjugate species in the other two compounds. Sodium ion doesn't appear on the table, but since NaOH is a strong base, its conjugate acid, sodium ion, must be a weak conjugate. From the chart, acetic acid is a stronger acid than ammonium ion. Therefore the correct order is

$$NaC_2H_3O_2 \quad < \quad NH_4Cl \quad < \quad HC_2H_3O_2 \quad \text{(most acidic)}$$

7. Nitric acid is a strong acid and so it is 100% ionized.

$$HNO_3 \quad \rightarrow \quad H^+ \quad + \quad NO_3^-$$

Thus, the concentration of hydrogen ion is equal to the initial concentration of HNO_3.

$$[H^+] = 0.50 \text{ M}$$

Solving for pH

$$pH = - \log[H^+]$$

$$= - \log(0.50)$$

$$\underline{pH = 0.30}$$

There are several ways to find pOH. We will use the relationship

$$pH \quad + \quad pOH = 14.00$$

$$pOH = 14.00 - pH \quad = 14.00 - 0.30$$

$$\underline{pOH = 13.70}$$

294

8. Since both H^+ and N_3^- are produced in equal molar amounts,

$$[N_3^-] = [H^+]$$

The $[H^+]$ is given, so

$$[N_3^-] = 1.38 \times 10^{-3} \text{ M}$$

The concentration of $[HN_3]$ that dissociated = $[H^+]$ formed, and so

$$[HN_3] = 0.10 - 1.38 \times 10^{-3}$$

The amount subtracted fails to make a difference in the last significant figure, and so may be ignored. Now place the values we have obtained in a data table.

	$[HN_3]$	$[H^+]$	$[N_3^-]$
equilibrium concentration	0.10	1.38×10^{-3}	1.38×10^{-3}

Write the equilibrium expression

$$K_a = \frac{[H^+][N_3^-]}{[HN_3]}$$

$$K_a = \frac{(1.38 \times 10^{-3})(1.38 \times 10^{-3})}{(0.10)}$$

$$\underline{K_a = 1.9 \times 10^{-5}}$$

9. Set up the data table for the data given

	$[(CH_3)_3N]$	$[(CH_3)_3NH^+]$	$[OH^-]$
Initial concentration	0.20	0	0
Change on going to equilibrium	-X	X	X
Equilibrium concentration	0.20 -X	X	X

Write the equilibrium expression,

$$K_b = \frac{[(CH_3)_3NH^+][OH^-]}{[(CH_3)_3N]}$$

Insert the values from the data table

$$7.4 \times 10^{-5} = \frac{x^2}{(0.20 - X)}$$

Test to determine if the X value can be ignored:

$$100 \times K_b = 7.4 \times 10^{-3}$$

Since 0.20 is greater than $100 \times K_b$, X can be ignored.

$$x^2 = 0.20 \times 7.4 \times 10^{-5}$$

$$X = [OH^-] = 3.8 \times 10^{-3} \text{ M}$$

$$pOH = -\log[OH^-] = -\log(3.8 \times 10^{-3})$$

$$pOH = 2.41$$

$$pH = 14.00 - pOH = 14.00 - 2.41$$

$$\underline{pH = 11.59}$$

10. a. H_2Se Lewis base   b. AsH_3 Lewis base

    c. Ag^+ Lewis acid   d. Ti^{4+} Lewis acid

11. Set up the data table as in the previous problem.

	[HF]	$[H^+]$	$[F^-]$
Initial concentration	0.050	0	0
Change on going to equilibrium	-X	+X	+X
Equilibrium concentration	0.050 - X	X	X

Next, write the equilibrium expression and insert the concentrations from the data table.

$$K_a = \frac{[H^+][F^-]}{[HF]}$$

$$7.2 \times 10^{-4} = \frac{X^2}{(0.050 - X)}$$

Notice that since 0.050 <u>is not greater</u> than 100 x K_a, X cannot be ignored! We must use the quadratic equation, which has the general form

$$aX^2 + bX - c = 0$$

After multiplying and rearranging our expression we obtain

$$X^2 + (7.2 \times 10^{-4})X - (3.6 \times 10^{-5}) = 0$$

By comparison with the standard quadratic form

$$a = 1; \quad b = 7.2 \times 10^{-4}, \text{ and } c = -3.6 \times 10^{-5}$$

Substitute these values into the equation

$$X = \frac{-b + \sqrt{b^2 - 4ac}}{2a}$$

$$X = \frac{-(7.2 \times 10^{-4}) + \sqrt{(7.2 \times 10^{-4})^2 - 4(-3.6 \times 10^{-5})}}{2 \cdot 1}$$

$$X = [H^+] = 5.65 \times 10^{-3} \text{ M}$$

If X had been ignored instead of using the quadratic, the answer obtained would have been $[H^+]$ = 6.0×10^{-3}, so the difference is significant.

Solving for the pH, using the equation pH = $-\log[H^+]$

$$pH = -\log(5.65 \times 10^{-3})$$

$$\underline{pH = 2.25}$$

12. When equal volumes of the two solutions are mixed, the concentrations of each of the two ions is decreased by half. The initial concentrations are therefore

$[Ag^+] = 1/2(0.010 \ M) = 0.0050 \ M$

$[Br^-] = 1/2(0.20 \ M) = 0.10 \ M$

Next, set up the data table

	$[Ag^+]$	$[Br^-]$	$[AgBr_2^-]$
Conc. after mixing	0.0050	0.10	0
After mixing and complete reaction	0	0.10 - 2(0.0050) =0.09	.0050

Next we imagine that the reaction reverses to a slight extent and obtain the equilibrium concentrations

	$[Ag^+]$	$[Br^-]$	$[AgBr_2^-]$
Conc. before equilibrium	0	0.10	0.0050
Change on going to equilibrium	+X	+2X	-X
Equilibrium concentration	X	0.09+2X	(.0050-X)

Write the equilibrium expression

$$K = \frac{[AgBr_2^-]}{[Ag^+][Br^-]^2}$$

and insert the equilibrium values in the equilibrium expression. The high value of the formation constant suggests that X will be extremely small compared to the concentrations of bromide ion and complex, so it can be ignored.

$$1.3 \times 10^7 = \frac{(0.0050)}{X(0.09)^2}$$

$$\underline{X = [Ag^+] = 4.7 \times 10^{-8} \ M}$$

The small value of X compared to the concentrations of bromide ion and silver complex ion shows that our assumption was reasonable.

PRACTICE TEST (45 min.)

1. For each reaction, identify the conjugate acid-base pairs and then use Table 17.2 from the textbook to predict whether the equilibrium lies predominantly to the left or to the right.

a. $H_2CO_3(aq) + CN^-(aq) \rightleftharpoons HCO_3^-(aq) + HCN(aq)$

b. $H_2O(aq) + NH_3(aq) \rightleftharpoons NH_2^-(aq) + H_3O^+(aq)$

2. a. The conjugate acid of HSO_4^- is _____.

   b. The conjugate base of HSO_4^- is _____.

3. The pOH of a 0.25 M aqueous solution of an unknown weak base, which we will designate BOH, is 3.10 at 25 degrees Celsius. What is the value of K_b for the equilibrium

$$BOH \rightleftharpoons B^+ + OH^-$$

4. Using Table 17.2 in the textbook, arrange these compounds in order of increasing strength of the conjugate base: HCl, $HC_2H_3O_2$, and HCN.

5. Calculate the pH of a 0.10 M solution of $Ba(OH)_2$

6. The value of K_w for water at 50°C is 5.48×10^{-14} M^2. What is the pH of a neutral aqueous solution at 50 degrees Celsius?

7. What is the pH of a 0.30 Molar solution of the salt NaF, if the hydrolysis of fluoride ion in water occurs according to the equation:

$$F^- + H_2O \rightleftharpoons HF + OH^- \qquad K_b = 1.4 \times 10^{-11}$$

8. At 25°C, a 0.200 M solution of a certain unknown

weak acid, HA, is 1.5% ionized. What is the K_a value for the equilibrium

$$HA \quad + \quad H_2O \quad \rightleftharpoons \quad H_3O^+ \quad + \quad A^-$$

9. Calculate the formation constant for the reaction

$$Ni^{2+}(aq) \quad + \quad 6 \ NH_3(aq) \quad \rightleftharpoons \quad [Ni(NH_3)_6]^{2+}$$

if a 0.20 M solution of this nickel complex ion is found to be 4.8% dissociated in aqueous solution.

10. Given that $K_a = 6.3 \times 10^{-5}$ for the reaction

$$C_6H_5COOH \quad \rightleftharpoons \quad C_6H_5COO^- \quad + \quad H^+$$

Calculate how many grams of benzoic acid must be dissolved in 150 milliliters of water in order to produce a solution having a pH of 2.50.

CONCEPT TEST ANSWERS

1. hydrogen ion, hydroxide ion
2. proton donor, proton acceptor
3. polyprotic, amphiprotic
4. $H_2PO_4^-$ and HPO_4^{2-} are one pair, H_2O and H_3O^+ are the other pair.
5. strong, strong
6. HNO_3 and HCl are strong acids; HOAc is a weak acid, and NaOH is a strong base.
7. If HF is a weak acid, F^- must be a strong conjugate base. Actually, HF is very corrosive, and by definition weak acids must be only about 1% ionized.
8. 14.00
9. hydronium ion, hydroxide ion
10. first ionization step
11. increases   12. increases
12. electron pair acceptor, electron pair donor
13. reciprocal
14. amphoteric

PRACTICE TEST ANSWERS

1. **(L1)** a. $H_2CO_3(aq) + CN^-(aq) \rightarrow HCO_3^-(aq) + HCN(aq)$

         $acid_1$ $base_2$ $base_1$ $acid_2$

Since H_2CO_3 is a stronger acid than HCN, the equilibrium goes to the right.

b. $H_2O(aq) + NH_3(aq) \rightarrow NH_2^-(aq) + H_2O^+(aq)$

     $base_1$ $acid_2$ $base_2$ $acid_1$

Since H_3O^+ is a stronger acid than NH_3, the equilibrium is shifted to the left.

2. **(L4)** a. H_2SO_4 b. SO_4^{2-}

3. **(L4)** 2.5×10^{-6}

4. **(L1)** (weakest) $Cl^- < C_2H_3O_2^- < CN^-$ (strongest)

5. **(L3)** pH = 13.3

6. **(L2)** pH = 6.63

7. **(L5)** pH = 8.31

8. **(L4)** $K_a = 4.6 \times 10^{-5}$ 9. **(L6)** 5.4×10^8

10. **(L4)** 2.9 grams

CHAPTER 18
REACTIONS BETWEEN
ACIDS AND BASES

LEARNING GOALS:

1. Be able to calculate the pH at the equivalence point in a titration involving a strong acid and a weak base, a weak acid and a strong base, or a strong acid and a strong base. (Sec. 18.1)

2. In order to follow the pH changes during a titration that involves a weak acid or base, it is essential to understand buffer solutions. In addition to simple acid-base titration, buffer solutions also have many other practical applications. (Sec. 18.2)

3. Be able to draw a simple graph showing the relationship between pH and volume of titrant in an acid-base titration and also be able to select a proper indicator based on the titration graph. (Sec. 18.3 & 18.4)

IMPORTANT NEW TERMS:

buffer (18.2)
common ion effect (18.2)

CONCEPT TEST

1. Acid-base reactions always proceed in the direction of the _____ (choose from stronger or weaker) acid-base pair.

2. Will the pH be greater than 7, less than 7, or equal to 7 at the equivalence point when a strong acid is titrated with a weak base? _____

3. When an acid and a base react in aqueous solution, one of the products is always a salt wherein the anion is the _____ of the acid and the cation is the _____ of the base.

4. Will the pH be greater than 7, less than 7, or equal to 7 at the equivalence point when sodium hydroxide is titrated with acetic acid? _____

5. Will an acidic or a basic solution result if equal molar amounts of a weak acid and a weak base are mixed, and the K_b of the base is larger than the K_a of the acid? _____

6. If sodium acetate, a salt, is added to a solution of acetic acid, a weak acid, will the pH of the resulting solution be higher, lower, or the same as that of the original acetic acid solution?

_____

7. The combination of a weak acid and its conjugate base or a weak base and its conjugate acid is used to prepare a special kind of solution called a(n) _____, because it will resist the change of pH when a strong acid or base is added.

8. When the color of the indicator changes during

an acid-base titration, it is most correctly described as the _____ of that titration rather than the equivalence point.

9. An indicator changes color when the hydrogen ion concentration is equal to the _____ of the indicator.

10. The defining equation for pK_a is

$$pK_a = \text{_____}.$$

11. In an acidic buffer solution,

$$[H^+] = \text{_____}.$$

12. When titrating a weak base with a strong acid, choose an indicator that changes at a pH of approximately _____.

STUDY HINTS:

1. One of the reasons that students find acid-base titration problems to be especially difficult is the fact that most problems really consist of three separate components. It is probably easiest to think of the process as involving three steps that must usually be accomplished in order to solve the problem.

    a. Determine the initial moles of acid and moles of base and then determine what acidic or basic species are present and how many moles of each is present. Be sure to identify each remaining species as a strong acid, strong base, weak acid, weak base, or salt. This information is critical in part c.

    b. Once the moles of each species present is known, it is necessary to calculate the concentration of each species. The final volume of the solution is the sum of the two (or more) components in the titration, and so be sure to use this new volume to determine the concentrations.

c. Based on the nature of the species remaining, determine whether the problem is a strong acid ionization, weak acid ionization, strong base ionization, weak base ionization, buffer, or hydrolysis problem. This should clearly identify the calculations necessary in the final step.

Breaking titration problems down into this three step process may make it easier to be successful with this type of problem.

2. There are some special cases that can greatly simplify the calculations in acid-base titrations. Learning to recognize these situations will sometimes save a great deal of work.

a. The pH is always 7 at the equivalence point for the titration of a strong acid and a strong base.

b. If a titration is at the end point, and if both reactants have the same concentration, the concentration of the products will be half as great as that of the original reactants.

Watch for examples of these situations in the problems in this chapter.

3. It should be obvious that the first step in solving a titration or buffer problem is to write the equation for the equilibrium. It is crucial that the correct equation be selected. In most cases it will simply consist of a weak acid or base (or a conjugate acid or base) ionizing in water. Don't try to make the equations too complicated. This is a case where the simplest approach is almost always the best.

4. Suppose it may be necessary to set up a problem involving the addition of a strong acid to a buffer consisting of a weak acid, HA, and its salt, NaA. The equilibrium for the buffer would be

$$HA \rightleftharpoons H^+ + A^-$$

Even though it might appear reasonable to add the concentration of the strong acid to the hydrogen ion that is already present, it is usually easier to work the problem by first allowing the acid to react with the conjugate base, A^-.

5. In the previous chapter, a rule of thumb was defined to determine whether X could be neglected when it was added to or subtracted from a much larger number. This rule used the equilibrium constant as a rough indication of the size of X. What would happen if the unknown, X, were an amount of reagent added to the solution separately rather than being the result of the equilibrium? Under these circumstances, it's no longer possible to use this rule of thumb as a justification for neglecting X. This type of problem is sometimes called a "Big X" problem, to emphasize that X cannot be neglected in the normal way. Be sure to carefully examine the examples of this type of problem that are included in both the practice problems and the practice test.

PRACTICE PROBLEMS

1. (L3) a. Draw a rough sketch of the titration graph that will result if KOH is added to HNO_3. b. Draw the titration graph that will result when KOH is added to formic acid, HCOOH.

2. (**L1**) Calculate the pH that would result if 25.0 mL of 0.200 M HNO_3 is added to 20.0 mL of 0.250 M KOH.

3. (**L1**) In a certain titration, 20.0 mL of 0.100 M hydrochloric acid is added to 50.0 mL of 0.150 M ammonia. What is the pH of the resulting solution? (K_b = 1.8x10^{-5} for NH_3)

4. (**L2**) What is the pH of a solution prepared by dissolving 13.4 grams of CH_3NH_3Cl in 250.0 mL of 0.210 M CH_3NH_2, methylamine. Assume there is no change in solution volume when the salt of methylamine is added. (K_b = 5.0x10^{-4} for CH_3NH_2)

5. (**L2**) How many grams of $NaC_2H_3O_2$ must be added to 1.00 liter of 0.200 M acetic acid in order to prepare a buffer having a pH of 5.00. Assume that there is no change in volume when the solid sodium acetate is added. ($K_a = 1.8x10^{-5}$ for acetic acid)

6. (**L2**) A buffer solution has been prepared by dissolving 0.500 moles of sodium formate, NaHCOO, in 1.00 liter of 0.500 Molar formic acid, $HCOOH$.
a. What is the pH of this solution? ($K_a = 1.8x10^{-4}$ for formic acid.)

b. Calculate the pH that will result if 0.010 mole of KOH is added to 1.00 L of the buffer solution described in part a.

7. **(L2)** The initial pH is 9.08 for a solution formed by mixing 0.300 moles of solid NH_4Cl into a 0.200 M solution of ammonia having a total volume of 1.00 liter. How many moles of HCl must be added to this solution to decrease the pH by 1.00 unit.

8. Calculate the pH at the equivalence point of a titration in which 50.0 mL of 0.300 M HF is added to 50.0 mL of 0.300 M KOH. ($K_b = 1.4 \times 10^{-11}$ for F^-.)

PRACTICE PROBLEM SOLUTIONS

1.

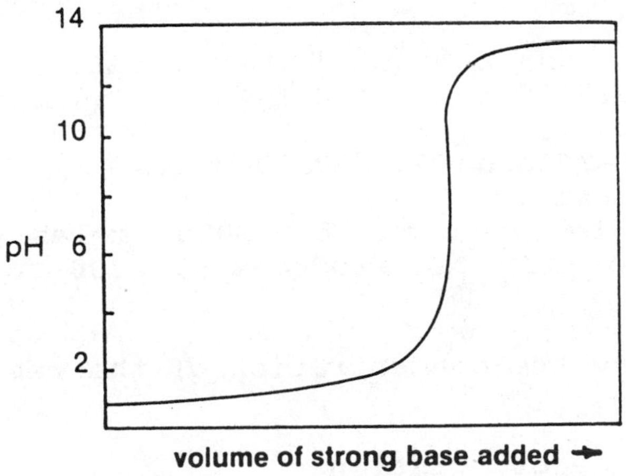

Figure 18.1 Titration curve for KOH and HNO_3

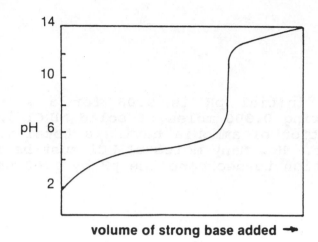

volume of strong base added →

Figure 18.2 Titration curve for the addition of KOH to HCOOH.

2. First, calculate the moles of acid and base.

mol of HNO_3 = 0.0250 liters x 0.200 M = 0.00500 mol
mol of KOH = 0.0200 liters x 0.250 M = 0.00500 mol

Moles of acid equals moles of base, and so the titration is at the equivalence point. Both HNO_3 and KOH are strong, and so there is no possibility of hydrolysis.

Since the acid and base are completely reacted, and there is no hydrolysis, the solution is neutral and pH = 7.
        The pH at the end point is 7.00

3. Calculate how many moles of each substance have been used.

Mol of HCl = 0.0200 L x 0.100 M = 0.00200 mol HCl
Mol of NH_3 = 0.0500 L x 0.150 M = 0.00750 mol NH_3

Moles of NH_3 remaining = 0.00550 moles

In addition, the reaction of 0.00200 moles of HCl and 0.00200 moles of NH_3 produces 0.00200 moles of NH_4Cl.

Next, determine the concentration of the remaining species.

The total volume of solution =
        0.0200 L + 0.0500 L = 0.0700 L

310

$[NH_3]$ = 0.00550 mol/0.0700 L = 0.07857 M

$[NH_4Cl]$ = 0.00200 mol/0.0700 L = 0.02857 M

Notice that a weak base and its salt are the remaining substances that will control the pH of the solution. The next step is to set up a data table and work a common ion problem using this information. The equilibrium reaction is

$$NH_3(aq) + H_2O(\ell) \rightleftarrows NH_4^+(aq) + OH^-(aq)$$

Next, prepare a data table.

	$[NH_3]$	$[NH_4^+]$	$[OH^-]$
Before ionization (M)	0.07857	0.02857	0
Change	$-X$	$+X$	$+X$
At equilibrium	0.07857$-X$	0.02857$+X$	X

Write the equilibrium expression, substitute the values from the data table, and solve.

$$K_b = \frac{[NH_4^+][OH^-]}{[NH_3]}$$

$$1.8 \times 10^{-5} = \frac{(0.02857+X)X}{(0.07857-X)}$$

According to the rule of thumb, X may be neglected

$$X = 4.95 \times 10^{-5} = [OH^-]$$

$$pOH = -log\ [OH^-] = -log\ (4.95 \times 10^{-5})$$

$$pOH = 4.31$$

$$pH = 14.00 - pOH = 14.00 - 4.31 =$$

$$pH = \underline{9.69}$$

4. Since the components are a weak base and a salt of the weak base, this is a common ion problem.

Begin by calculating the concentration of the CH_3NH_3Cl.

Moles CH_3NH_3Cl = 13.4 g/67.45 g/mole = 0.1987 mol

Molarity CH_3NH_3Cl = 0.1987 mol/0.250 L = 0.7947 M

Based on the following equilibrium reaction

$$CH_3NH_2(aq) + H_2O(\ell) \rightleftharpoons CH_3NH_3^+(aq) + OH^-(aq)$$

Next, write the data table.

	$[CH_3NH_2]$	$[CH_3NH_3^+]$	$[OH^-]$
Before dissociation (M)	0.210	0.7947	0
Change	-X	+X	+X
Equilibrium concentrations	0.210-X	0.7947+X	X

Substituting into the equilibrium expression

$$K_b = \frac{[CH_3NH_3^+][OH^-]}{[CH_3NH_2]}$$

$$5.0 \times 10^{-4} = \frac{(0.7947+X)X}{(0.210-X)}$$

According to the rule of thumb, X may be neglected.

$$X = 1.32 \times 10^{-4} \text{ M} = [OH^-]$$

$$pOH = -\log[OH^-] = -\log(1.32 \times 10^{-4})$$

$$pOH = 3.88$$

$$pH = 14.00 - pOH = 14.00 - 3.88$$

$$pH = \underline{10.12}$$

5. The components of this solution are a weak acid and its salt, so this is a buffer problem. Begin by writing the balanced equation for the reaction.

$$HC_2C_3O_2(aq) \rightleftharpoons H^+(aq) + C_2H_3O_2^-(aq)$$

The unknown quantity in this problem is the mass of sodium acetate added. This value can be designated to be X, but since the relationship between the magnitude of X and magnitude of the equilibrium constant is not simple, it will not be possible to use the rule of thumb and decide to neglect this X. This is an example of a "Big X" problem of the type discussed in the study hints section.

It's given that the pH will be 5.00, therefore

$$pH = 5.00 = -log\ [H^+]$$

$$[H^+] = 1.0 \times 10^{-5}$$

Now develop a data table using this information.

	$[HC_2H_3O_2]$	$[H^+]$	$[C_2H_3O_2^-]$
Initial concs.	0.200	0	X
Change when proceeding to equilibrium	-1.0×10^{-5}	$+1.0 \times 10^{-5}$	1.0×10^{-5}
Equilibrium concs.	$(0.200-1.0 \times 10^{-5})$	1.0×10^{-5}	$(X+1.0 \times 10^{-5})$

Write the equilibrium expression and substitute the equilibrium concentrations from the data table.

$$Ka = \frac{[C_2H_3O_2^-][H^+]}{[HC_2H_3O_2]}$$

$$1.8 \times 10^{-5} = \frac{(X + 1.0 \times 10^{-5})(1.0 \times 10^{-5})}{(0.200 - 1.0 \times 10^{-5})}$$

The concentration change when proceeding to equilibrium is obviously small when compared with the initial concentration of $HC_2H_3O_2$ and is probably small when compared with the concentration of $NaC_2H_3O_2$ added. For the time being, this value can be neglected, but this assumption must be checked when the problem is completed.

Substituting into the equilibrium expression

$$1.8 \times 10^{-5} = \frac{X(1.0 \times 10^{-5})}{(0.200)}$$

$$\underline{X = 0.36 \text{ M} = \text{concentration of } NaC_2H_3O_2 \text{ added}}$$

Notice that 1.0×10^{-5} is negligible compared with X.

To calculate how many grams of sodium acetate are needed, determine the number of moles and multiply this by the molar mass of sodium acetate.

grams of $NaC_2H_3O_2$ = 0.36 M x 1.00 L x 82.0 g/mol

$$\underline{\text{grams of } NaC_2H_3O_2 = 29.5 \text{ grams}}$$

6. a. Begin by writing the balanced equation.

$$HCOOH(aq) \rightleftharpoons H^+(aq) + HCOO^-(aq)$$

and then set up the table of concentrations.

	[HCOOH]	[H⁺]	[HCOO⁻]
Initial concs. (M)	0.500	0	0.500
Change on proceeding to equilibrium	-X	+X	+X
Equilibrium concs. (M)	0.500-X	X	0.500+X

Substitute these values into the equilibrium expression.

$$Ka = 1.8 \times 10^{-4} = \frac{[H^+][HCOO^-]}{[HCOOH]} = \frac{X(0.500+X)}{(0.500-X)}$$

314

Solving for hydrogen ion concentration with the assumption that X may be neglected

$$X = [H^+] = 1.8 \times 10^{-4}$$

$$pH = -\log[H^+] = -\log(1.8 \times 10^{-4})$$

$$pH = 3.74$$

b. The equilibrium remains the same, but in this case 0.010 moles of KOH has been added. Write the data table again, but this time allow the KOH to react with the HCOOH, decreasing the HCOOH concentration and producing more HCOO$^-$ ion.

The appropriate data table is

	[HCOOH]	[H$^+$]	[HCOO$^-$]
Initial concs. (M)	0.500	0	0.500
Change on adding KOH	−0.010		+0.010
Concentration after adding KOH	0.490		0.510
Change on proceeding to equilibrium	−X	+X	+X
Equilibrium concs. (M)	0.490−X	X	0.510+X

Substituting into the equilibrium expression

$$K_a = 1.8 \times 10^{-4} = \frac{[H^+][HCOO^-]}{[HCOOH]} = \frac{X(0.510+X)}{(0.490-X)}$$

Neglecting X and solving

$$X = [H^+] = 1.73 \times 10^{-4}$$

$$pH = -\log[H^+] = -\log(1.73 \times 10^{-4})$$

$$pH = 3.76$$

As expected, the buffer solution shows a very small change in pH despite the addition of a strong base.

7. This is a buffer solution, and the relationship among the various species in solution is given by the equation

$$NH_3(aq) + H_2O(\ell) \rightleftarrows NH_4^+(aq) + OH^-(aq)$$

When HCl, a strong acid, is added, it will react with the conjugate base, NH_3, and the data table will show this change. The amount of HCl added is an unknown, X, but since it is not directly related to the equilibrium constant, it's not possible to neglect X. This is an example of a "Big X" problem of the type described in the Hints.

Given that the final pH is 8.08

$$[OH^-] = antilog(- pOH) = antilog (-5.92)$$

$$[OH^-] = 1.202 \times 10^{-6} M.$$

Now write the data table.

	$[NH_3]$	$[NH_4^+]$	$[OH^-]$
Initial concs.	0.200	0.300	0
Change due to HCl added	-X	+X	0
Concs. after HCl added	0.200-X	0.300+X	0
Change when proceeding to equilibrium	-1.202×10^{-6}	$+1.202 \times 10^{-6}$	$+1.202 \times 10^{-6}$
Equilibrium concs.	0.200-X- 1.202×10^{-6}	0.300+X+ 1.202×10^{-6}	1.202×10^{-6}

It appears that 1.202×10^{-6}, the change upon proceeding to equilibrium is small enough so that its effect on the concentration of NH_3 and NH_4^+ can be ignored, but this assumption will need to be checked when the problem is completed.

Now write the equilibrium expression and solve using the values from the data table.

$$K_b = \frac{[NH_4^+][OH^-]}{[NH_3]}$$

$$1.8 \times 10^{-5} = \frac{(0.300+X)(1.202 \times 10^{-6})}{(0.200-X)}$$

X = 0.169 M = concentration of HCl added

(Notice that the decision to neglect 1.202×10^{-6} was justified.)

Since the solution has a volume of 1.00 L

Moles of HCl = 0.169 M x 1.00 L

Moles of HCl = 0.169 moles

8. Determine the moles of each reactant present.

Moles of HF = 0.0500 L x 0.300 M = 0.0150 moles
Moles of KOH = 0.0500 L x 0.300 M = 0.0150 moles

The titration is at the equivalence point, and acidity of the solution will be determined by the hydrolysis of the salt, NaF.

Conc. of NaF = 0.0150 moles/0.100 L = 0.150 M

The F^- ion is the strong conjugate, and the equation for its hydrolysis is

F^-(aq) + H_2O(ℓ) ⇌ HF(aq) + OH^-(aq)

Now set up the data table for this situation.

	$[F^-]$	[HF]	$[OH^-]$
Before hydrolysis (M)	0.150	0	0
Change on proceeding to equilibrium	-X	+X	+X
Equilibrium conc. (M)	(0.150-X)	X	X

Inserting these values into the the equilibrium expression

$$K_b = \frac{[HF][OH^-]}{[F^-]} = \frac{x^2}{(0.150-X)} = 1.4 \times 10^{-11}$$

The rule of thumb indicates that X may be neglected.

$$X = [OH^-] = 1.45 \times 10^{-6}$$

$$pOH = -\log[OH^-] = -\log(1.45 \times 10^{-6}) = 5.84$$

$$pH = 14.00 - pOH = 14.00 - 5.84$$

$$\underline{pH = 8.16}$$

PRACTICE TEST (60 min.)

1. Calculate the pH that results when 0.25 moles of sodium acetate is dissolved in 500 mL of 0.15 M acetic acid. ($K_a = 1.8 \times 10^{-5}$ for acetic acid. Assume no volume change upon addition of solid.)

2. Calculate the pH that would result if 25.0 mL of 0.150 M formic acid, HCOOH, is added to 25.0 mL of 0.200 M NaOH. ($K_a = 1.8 \times 10^{-4}$ for formic acid.)

3. Calculate the pH at the equivalence point of a titration involving 0.30 M NH_3 and 0.30 M HCl. ($K_a = 5.6 \times 10^{-10}$ for ammonium ion. Notice that the pH will not depend on the volumes of acid and base involved.)

4. How many moles of solid NH_4Cl must be added to 1.00 liter of 0.200 NH_3 in order to produce a final solution having a pH of 9.40? (Assume that the addition of the NH_4Cl doesn't change the volume of the solution. $K_b = 1.8 \times 10^{-5}$ for ammonia.)

5. Calculate the pH that results when 50.0 mL of 0.200 M KOH is added to 100.0 mL of 0.200 M HF. ($K_a = 7.2 \times 10^{-4}$ for hydrofluoric acid.)

6. Solid sodium acetate is added to 500.0 mL of 0.300 M $HC_2H_3O_2$ until the pH of the resulting solution is 5.70 at 25°C. Assuming that there is no volume change when the solid is added, how many grams of sodium acetate were added? ($K_a = 1.8 \times 10^{-5}$ for acetic acid.)

7. Which of the following pairs of compounds below would be most appropriate for the preparation of a buffer having a pH of 5.0? Suggest what concentration of each compound might be used to produce exactly this pH.

a. C_5H_5N/C_5H_5NHCl $K_b = 1.7 \times 10^{-9}$

b. NH_3/NH_4Cl $K_b = 1.8 \times 10^{-5}$

c. $HC_2H_3O_2/NaC_2H_3O_2$ $K_a = 1.8 \times 10^{-5}$

d. HF/NaF $K_a = 6.8 \times 10^{-4}$

8. A balanced buffer is prepared by dissolving 1.00 mole of $NaC_2H_3O_2$ in 1.00 liter of 1.00 M $HC_2H_3O_2$ and is found to have a pH of 4.74. How many moles of solid NaOH must be added to this solution to increase the final pH to 5.75?

CONCEPT TEST ANSWERS

1. weaker   2. less than 7
3. conjugate base, conjugate acid
4. greater than 7   5. basic
6. higher pH (less acidic)
7. buffer   8. end point
9. K_a 10. $-\log(K_a)$

11. $\dfrac{K_a \times [acid]}{[conjugate\ base]}$ 12. eight

PRACTICE TEST ANSWERS

1. (**L2**) pH = 5.27
2. (**L2**) pH = 12.40
3. (**L1**) 5.04
4. (**L2**) 0.14 mole

5. **(L1)** 3.17
6. **(L2)** 111 grams
7. **(L2)** $HC_2H_3O_2/NaC_2H_3O_2$, as long as $[C_2H_3O_2^-]$ = 1.8$[HC_2H_3O_2]$, the correct pH will result.
8. **(L2)** 0.82 moles

CHAPTER 19
PRECIPITATION REACTIONS

LEARNING GOALS:

1. Based on the equations for precipitation reactions, write the solubility product expressions for insoluble salts. (Sec. 19.2)

2. Be able to calculate the solubility product constant based on solubility information, or if given the value of the solubility product constant for a compound, be able to calculate the solubility and/or concentration of the ions produced by that compound in aqueous solution. It's also important to determine which of two salts is more soluble, based on the K_{sp} values. (Sec. 19.3 & 19.4)

3. When given the value of the solubility product for a compound and the concentrations of the component ions in solution, be able to predict whether or not a precipitate will form. This technique important when using solubility differences to accomplish the separations of metal ions. (Sec. 19.5)

4. The presence of a common ion will shift the position of a solubility equilibrium just as was observed with other equilibria. The method of solution is very similar to problem types encountered previously, so this problem type should be relatively familiar. (Sec. 19.6)

5. Be able to evaluate the probable effectiveness of proposed methods for selective precipitation and also to develop new selective precipitation procedures based on solubilities or solubility product values. (Sec. 19.7)

6. Review the description of the formation of complex ions (Chap. 17) in order to use this information to predict the effects of simultaneous equilibrium involving both solubility and complex formation. (Sec. 19.8 & 19.9)

IMPORTANT NEW TERMS:

ion separation (19.7)
qualitative analysis (19.7)
selective precipitation (19.10)
simultaneous equilibria (19.8)
solubility product (19.2)

CONCEPT TEST

1. A precipitation reaction is an exchange reaction in which one of the products is _____.

2. When the salt MgF_2 is dissolved in water, the resulting concentration of fluoride ion will be twice as great as that of the magnesium ion. Will this relationship always be true?

_____

3. When a common ion is added to a salt solution, the solubility of the salt _____. (Choose from increases, decreases, or remains the same).

4. If carbonate ion is slowly added to a solution that is 0.01 M in both Ba^{2+} and Ca^{2+} ions, based on Table 19.2 in the textbook, which will precipitate first, $CaCO_3$ or $BaCO_3$? _____

5. In order to increase the solubility of $Ca(OH)_2$, is it better to use a strong acid or a strong base as the solvent? _____

6. The text suggests that insoluble compounds will usually dissolve to the extent of less than _____ mole per liter.

7. Calcite is more soluble in the cold water of the deep oceans than it is in the warmer surface waters. Does this suggest that dissolving calcite

is an exothermic or an endothermic process?
_____.

8. If the value of the reaction quotient, Q, is greater than the value of K_{sp}, the system is said to be _____.

9. If an anion is added to a solution containing equal concentrations of two cations that form insoluble salts with that anion, the substance that precipitates first will be the one whose _____ is exceeded first.

10. Insoluble inorganic salts containing _____ _____ tend to be soluble in solutions of strong acids.

11. Any salt will dissolve in water to a greater extent than given by K_{sp} if it contains an anion that is _____.

STUDY HINTS:

1. When working common ion problems, students sometimes become confused about when to use the stoichiometric coefficients. To clarify this question, examine where the ion came from. For example, consider the addition of NaCl to a solution of $MgCl_2$. The chloride ion that results from the ionization of the magnesium chloride will be twice as great as the concentration of the dissolved magnesium chloride, but the chloride ion from the sodium chloride will be equal to the concentration of dissolved sodium chloride. Of course, if it is stated that the chloride ion concentration has a certain value, it is not necessary to multiply regardless of the source of the chloride ion.

2. For some reason, when the same stoichiometric factor is used to both multiply a concentration and raise to a power, some students are more likely to omit one or the other of these steps. Remember

that the same stoichiometric factor may need to be used both in the concentrations for the data table as well as in the solubility product expression!

3. In the previous chapter, it was noted that in some cases it isn't possible to use the rule of thumb to neglect X. A similar situation is also found with solubility problems, so watch carefully for these "Big X" problems

PRACTICE PROBLEMS

1. (**L1**) Write the K_{sp} expression for each of the following salts:

a. AgI

b. $Cr(OH)_3$

c. Ag_3PO_4

d. Co_2S_3

2. (**L2**) The solubility of AgCl is 1.3×10^{-5} M. What is the value of K_{sp} for this compound?

3. (**L2**) Given that $K_{sp} = 6.4 \times 10^{-9}$ for MgF_2, a. calculate the solubility of MgF_2 in water in mol/L. b. calculate the solubility of MgF_2 in water in mg/L

4. Will a precipitate form if 0.00010 moles of solid NaCl is added to 500 milliliters of 0.0020 M $Pb(NO_3)_2$? ($K_{sp} = 1.7 \times 10^{-5}$ for $PbCl_2$.)

5. (**L4**) Calculate the solubility (mol/L) of PbI_2 in a 0.20 Molar solution of NaI. ($K_{sp} = 8.7 \times 10^{-9}$)

6. (**L4**) Calculate the pH required in an aqueous solution in order to reduce the solubility of $Al(OH)_3$ to 2.0×10^{-14} M. ($K_{sp} = 1.9 \times 10^{-33}$ for $Al(OH)_3$

7. **(L5)** If hydroxide ion is very slowly added to a solution containing 0.0010 M Fe^{2+} and 0.0010 M Co^{2+}, which ion will precipitate first, and what is the concentration of the first ion that remains in solution when the second ion just begins to precipitate? ($K_{sp} = 7.9 \times 10^{-15}$ for $Fe(OH)_2$ and $K_{sp} = 2.5 \times 10^{-16}$ for $Co(OH)_2$.)

PRACTICE PROBLEM SOLUTIONS

1. a. $K_{sp} = [Ag^+][I^-]$

   b. $K_{sp} = [Cr^{3+}][OH^-]^3$

   c. $K_{sp} = [Ag^+]^3[PO_4^{3-}]$

   d. $K_{sp} = [Co^{3+}]^2[S^{2-}]^3$

2. When silver chloride dissolves in water, the reaction can be represented

$$AgCl(s) \rightleftharpoons Ag^+(aq) + Cl^-(aq)$$

and the K_{sp} expression for AgCl is

$$K_{sp} = [Ag^+][Cl^-]$$

Each mole of AgCl that dissolves will produce one mole of Ag^+ and one mole of Cl^-, so if 1.3×10^{-5} mol/L dissolves then

$$[Ag^+] = [Cl^-] = 1.3 \times 10^{-5}$$

326

Substituting into the K_{sp} expression

$$K_{sp} = (1.3 \times 10^{-5})^2$$

$$\underline{K_{sp} = 1.7 \times 10^{-10}}$$

3. a. The equation for the dissolving of MgF_2 is

$$MgF_2(s) \rightleftarrows Mg^{2+}(aq) + 2 F^-(aq)$$

and the K_{sp} expression is

$$K_{sp} = [Mg^{2+}][F^-]^2$$

If S = the solubility of this salt,
then $S = [Mg^{2+}] = 2[F^-]$.

Substituting these values into the K_{sp} expression

$$K_{sp} = [Mg^{2+}][F^-]^2$$

$$K_{sp} = (S)(2S)^2 = 4S^3 = 6.4 \times 10^{-9}$$

$$\underline{S = 1.2 \times 10^{-3} \text{ M}}$$

b. Now convert the solubility in mol/L to mg/L.

$$S(mg/L) = 1.2 \times 10^{-3} \text{ mol/L} \times 62.3 \text{ g/mol} \times 1000 \text{ mg/g}$$

$$\underline{S(mg/L) = 73 \text{ mg/L}}$$

4. The equation for the dissolving of the salt is

$$PbCl_2(s) \rightleftarrows Pb^{2+} + 2Cl^-$$

And so the K_{sp} expression for $PbCl_2$ is

$$K_{sp} = [Pb^{2+}][Cl^-]^2$$

Since NaCl and $Pb(NO_3)_2$ are quite soluble, the concentration of the Cl^- and Pb^{2+} ions is equal to the concentrations of the salts.

$$[Pb^{2+}] = 0.0020 \text{ M}$$

Notice that the chloride ion concentration must be calculated from the moles of compound and volume of solution.

$$[Cl^-] = 0.00010 \text{ mol}/0.500 \text{ L} = 0.00020 \text{ M}$$

[Notice that the source of the chloride ion in this case was NaCl, and so it was not necessary to double the concentration. If the chloride ion had been produced by the $PbCl_2$, the situation would have been different.]

Now substitute these values into the K_{sp} expression.

$$Q = (0.0020)(0.00020)^2 = 8.0 \times 10^{-11}$$

Since $Q < 1.7 \times 10^{-5}$, a precipitate will not form.

5. Sodium iodide is quite soluble, so assume that all of it is in solution and completely dissociated. By definition, determining the solubility of PbI_2 indicates that some solid doesn't dissolve.

Write the equations for the two compounds in water.

$$NaI(s) \longrightarrow Na^+(aq) + I^-(aq)$$

$$PbI_2(s) \rightleftarrows Pb^{2+}(aq) + 2 I^-(aq)$$

Set up a data table for this solution, using S to represent the solubility of the lead iodide:

	$[Pb^{2+}]$	$[I^-]$
Initial concentration (M)	0	0.20
Change on proceeding to equilibrium	+S	+2S
Equilibrium concentration	S	0.20+2S

[Notice that the iodide concentration that resulted from the ionization of the NaI is not doubled, but the iodide concentration that results from the dissolving of the lead iodide is doubled. Examination of the equations for these reactions should make it obvious why this is true.]

Now write the K_{sp} expression and substitute these values.

$$K_{sp} = 8.7 \times 10^{-9} = [Pb^{2+}][Cl^-]^2 = S(0.20+2S)^2$$

The value of S is quite small, even in pure water, and the result of the common ion will be to decrease this concentration even further. It is possible to assume that 0.20+2S is approximately equal to 0.20 without causing a significant error.

The equation then becomes

$$K_{sp} = (0.20)^2 S = 8.7 \times 10^{-9}$$

$$\underline{S = 2.2 \times 10^{-7} \text{ M}}$$

Checking the assumption, it is clear that 0.20 is much larger than the value of 2S.

6. Begin by writing the equation for the dissolving of the aluminum hydroxide.

$$Al(OH)_3(s) \rightleftharpoons Al^{3+}(aq) + 3\ OH^-(aq)$$

The question suggests that a base is being added to increase the pH and decrease the solubility of the $Al(OH)_3$. The hydroxide ion concentration will be a combination of the solubility of the aluminum hydroxide and this other source of base. The best way to show this is with a data table.

	$[Al^{3+}]$	$[OH^-]$
Equilibrium concs. (M)	S	3S+X

The required solubility is 2.0×10^{-14} M, and this value can be substituted for S in the data table to produce:

Equilibrium concs. (M)	2.0×10^{-14}	$X+3(2.0 \times 10^{-14})$

329

Notice that this is another case where the X value cannot be neglected. Usually the additional concentration of common ion is rather large compared to the result of the dissolving of the compound, and so now assume that X >> 2.0×10^{-14}. These values can now be substituted into the K_{sp} expression.

$$K_{sp} = 1.9 \times 10^{-33} = [Al^{3+}][OH^-]^3 = (2.0 \times 10^{-14})X^3$$

$$X = 4.6 \times 10^{-7} = [OH^-]$$

$$pOH = -\log[OH^-] = -\log(4.6 \times 10^{-7}) = 6.34$$

$$\underline{pH = 7.66}$$

7. Calculate the concentration of hydroxide ion necessary to precipitate each of these ions.

For Fe(OH)$_2$,
$$K_{sp} = [Fe^{2+}][OH^-]^2$$

and so
$$[OH^-]^2 = \frac{K_{sp}}{[Fe^{2+}]} = \frac{7.9 \times 10^{-15}}{0.0010}$$

$$[OH^-] = 2.8 \times 10^{-6}$$

For Co(OH)$_2$,
$$K_{sp} = [Co^{2+}][OH^-]^2$$

or
$$[OH^-]^2 = \frac{K_{sp}}{[Co^{2+}]} = \frac{2.5 \times 10^{-16}}{0.0010}$$

$$[OH^-] = 5.0 \times 10^{-7}$$

<u>Since the cobalt(II) requires a smaller concentration of OH⁻ it will be the first ion to precipitate.</u>

The next question really asks what concentration of cobalt(II) will remain in solution when the iron(II) just begins to precipitate. It has already been shown that iron(II) will just begin to precipitate when $[OH^-] = 2.8 \times 10^{-6}$. How much cobalt(II) will remain in solution when the hydroxide concentration reaches that level?

Substitute the concentration of hydroxide ion when the iron just begins to precipitate, into the solubility product expression for the cobalt hydroxide to find the concentration of cobalt ion remaining at that time.

$$[Co^{2+}] \text{ as } Fe^{2+} \text{ just begins to precipitate} = \frac{K_{sp}}{[OH^-]^2}$$

$$[Co^{2+}] \text{ remaining} = \frac{2.5 \times 10^{-16}}{(2.8 \times 10^{-6})^2} = 3.2 \times 10^{-5} \text{ M}$$

Thus, when iron(II) hydroxide begins to precipitate

$$[Co^{2+}] = 3.2 \times 10^{-5} \text{ M}$$

PRACTICE TEST (45 min.)

1. Arrange the following compounds in order of decreasing solubility in water:

PbS	$K_{sp} = 8.4 \times 10^{-28}$
CdS	$K_{sp} = 3.6 \times 10^{-29}$
NiS	$K_{sp} = 3.0 \times 10^{-21}$

2. If $K_{sp} = 6.3 \times 10^{-6}$ for $PbBr_2$ in aqueous solution, calculate the solubility of this salt in water.

3. Given that $K_{sp} = 1.8 \times 10^{-10}$ for AgCl, determine whether or not a precipitate will form if 0.00010 moles of solid NaCl is added to 500 milliliters of 0.0020 M $AgNO_3$.

4. If $K_{sp} = 8.1 \times 10^{-19}$ for BiI_3, what is the solubility of this salt in water?

5. If $K_{sp} = 3.9 \times 10^{-11}$ for CaF_2, what is the Molar solubility of CaF_2 in a 0.20 Molar $CaCl_2$ solution?

6. Calculate the pH required in an aqueous solution in order to reduce the solubility of $Mg(OH)_2$ to 2.0×10^{-6} M. ($K_{sp} = 1.5 \times 10^{-11}$ for $Mg(OH)_2$)

7. If sulfide ion is very slowly added to a solution containing 0.0050 M Zn^{2+} and 0.0050 M Co^{2+}, which ion will precipitate first, and what is the concentration of this ion that remains in solution when the second ion just begins to precipitate? ($K_{sp} = 5.9 \times 10^{-21}$ for CoS and $K_{sp} = 1.1 \times 10^{-21}$ for ZnS.)

8. Based on the equilibrium constant values for the following reactions:

$$Cd(NH_3)_4^{2+}(aq) \rightleftharpoons Cd^{2+}(aq) + 4\ NH_3(aq) \quad K = 1.0 \times 10^{-7}$$

$$CdS(s) \rightleftharpoons Cd^{2+}(aq) + S^{2-}(aq) \quad K = 3.6 \times 10^{-29}$$

Calculate the value of the equilibrium constant for the reaction

$$CdS(s) + 4\ NH_3(aq) \rightleftharpoons Cd(NH_3)_4^{2+}(aq) + S^{2-}(aq)$$

CONCEPT TEST ANSWERS

1. an insoluble compound
2. Yes, unless another compound is present that can also serve as a source of magnesium or fluoride ions. For example, if the solution contains both MgF_2 and NaF, the concentration of fluoride will not be twice that of the magnesium ion.
3. decreases
4. $CaCO_3$
5. strong acid
6. 0.01 mole
7. exothermic
8. supersaturated
9. K_{sp}
10. anions derived from weak bases
11. conjugate base of a weak acid

PRACTICE TEST ANSWERS

1. (L2) NiS > PbS > CdS
2. (L1) 1.2×10^{-2} M
3. (L3) Yes, since $Q = 4 \times 10^{-7}$, which is greater than

K_{sp}, a precipitate will form .

4. **(L1)** 1.3×10^{-5} 5. **(L4)** 7.0×10^{-6}
6. **(L5)** 11.44
7. **(L1)** ZnS will precipitate at a lower sulfide ion concentration (i.e. 2.2×10^{-19}) and when CoS is about to begin to precipitate the concentration of zinc remaining in solution is 9.2×10^{-4} M
8. **(L6)** 3.6×10^{-22}

CHAPTER 20
THE SPONTANEITY OF CHEMICAL REACTIONS:
ENTROPY AND FREE ENERGY

LEARNING GOALS:

1. As noted previously, energy change plays an important role in determining the spontaneity of a reaction. This chapter introduces a new state function, entropy, which is another fundamental measure of reaction spontaneity. Be able to qualitatively predict whether entropy increases or decreases for simple processes and also be able to calculate the entropy change for changes of state and chemical reactions. (Sec. 20.3 & 20.4)

2. This chapter also introduces the third major state function that determines reaction spontaneity, Gibbs free energy. Be able to calculate free energy changes using a table of standard free energy values and the equation

$$\Delta G^o_f = \Sigma m \Delta G^o_f (\text{products}) - \Sigma n \Delta G^o_f (\text{reactants})$$

Also understand the relationship between the sign of the free energy change and reaction spontaneity. (Sec. 20.5)

3. Given the standard enthalpy and entropy changes for a process, know how to use the formula

$$\Delta G^o = \Delta H^o - T \Delta S^o$$

to determine the free energy change at any temperature. (Sec. 20.5)

4. Gibbs free energy is closely related to the equilibrium constant by the equation

$$\Delta G^o = - RT \ln K$$

Be able to determine the K value, given appropriate

thermodynamic values, such as entropy change, enthalpy change, or free energy change. (Sec. 20.6)

IMPORTANT NEW TERMS:

absolute entropy (20.3)
entropy (20.3)
Gibbs free energy (or free energy) (20.5)
second law of thermodynamics (20.5)
spontaneous chemical reaction (20.1)
standard molar free energy of formation (20.5)
standard state entropy (20.3)
third law of thermodynamics (20.3)

CONCEPT TEST

1. A chemical reaction that, given sufficient time, achieves chemical equilibrium by reacting from left to right as written is said to be _____.

2. What two factors must be considered when trying to predict whether or not a particular process will be spontaneous?

a. _____

b. _____

3. The thermodynamic function called _____ is a measure of the disorder of a system.

4. The entropy of a pure, perfectly crystalline substance at zero kelvin is _____. This statement is called the _____.

5. Indicate whether each process below represents an increase or a decrease in entropy of the system.

a. Water evaporates from a beaker

b. Sugar dissolves in coffee.

c. Atmospheric oxygen dissolves in lake waters

d. An ice cube melts.

6. The second law of thermodynamics states that the

_____ of a system and its surroundings always increases in a spontaneous process.

7. For equilibrium processes, ΔS must equal _____.

8. In order for a process to be spontaneous, the sign of the free energy change must be _____ (choose from positive, negative, or zero.)

9. Elements in their standard states have free energy change values of _____.

10. In the relationship, $\Delta G° = - RT \ln K$, _____ is the value normally used for R.

STUDY HINTS:

1. Study sections 20.3 and 20.4 of the textbook carefully. They provide some valuable hints for understanding entropy better and predicting the sign of the entropy change for simple processes.

2. When working with quantities like $\Delta G°$ or $\Delta H°$, remember that the superscript automatically indicates the temperature must be 25°C, even if this value is not listed in the problem.

3. In a number of cases, it's a general rule that the value of certain thermodynamic variables is always zero for an element in it's standard state under standard conditions. Remember that this is not usually true for the absolute entropy values of the elements.

PRACTICE PROBLEMS

1. (L1) By inspection, predict whether the entropy change will be increase or decrease for the following:

a. $H_2O(\ell)$ → $H_2O(s)$

b. $O_2(g)$ → $2\ O(g)$

c. $NaCl(s)$ → $NaCl(aq)$

d. $NH_3(g)$ → $NH_3(aq)$

2. (**L2**) The standard free energy of formation for diamond is listed as 2.90 kJ/mole. Which is more stable, diamond or graphite? Does this agree with the common idea regarding these two substances?

3. (**L1**) The enthalpy of vaporization of carbon tetrachloride is 30.0 kJ/mole at it normal boiling point of 76.8°C. Calculate the entropy change that occurs when 1.0 mole of liquid carbon tetrachloride vaporizes at the boiling point.

4. (**L2**) Hydrazine, N_2H_4, and its methyl derivatives are used extensively for rocket fuels in guided missiles, space shuttles, and lunar landers. It can be used with a variety of oxidizers, including O_2, H_2O_2, and F_2. The equation for one such reaction is

$$N_2H_4(\ell) \ + \ 2 \ H_2O_2(\ell) \ \rightarrow \ N_2(g) \ + \ 4 \ H_2O(g)$$

Using the free energy values provided below, calculate the standard free energy change for this reaction.

Standard Free Energies of Formation (kJ/mol)

$N_2H_4(\ell)$ 149.2; $H_2O_2(\ell)$ -120.4; $H_2O(g)$ -228.6

5. (**L3**) As shown in the equation below, nitrogen oxide, NO, is a common pollutant that is produced by the reaction of nitrogen and oxygen gases. Since these gases are the major components of air, nitrogen oxide results when air is heated in furnaces, motor engine cylinders, or other high temperature combustion areas.

$$N_2(g) \ + \ O_2(g) \ \rightleftarrows \ 2 \ NO(g)$$

a. Given that the standard enthalpy of formation for NO(g) is 90.25 kJ/mole and using the absolute entropies listed below, calculate the free energy change for this reaction at 25°C. Is this reaction is spontaneous at 25°C?

Absolute Entropies
$N_2(g)$	191.5 J/mol·K	$O_2(g)$	205.0 J/mole·K
NO(g)	210.7 J/mol·K		

b. If the reaction is not spontaneous at 25°C, determine under what conditions it would be spontaneous.

6. **(L4)** Given that ΔG° = -100.4 kJ/mol for the reaction

$$CCl_4(\ell) + H_2(g) \rightleftharpoons HCl(g) + CHCl_3(g)$$

calculate the value of the equilibrium constant, K, for this reaction at 25°C.

7. When sulfur-containing fuels, such as coal or fuel oil, are burned, one of the products is SO_2, which is converted to SO_3 in the air according to the equation

$$2\ SO_2(g) + O_2(g) \rightleftharpoons 2\ SO_3(g)$$

Given the thermodynamic data in the table below, calculate K for this reaction at 25°C.

	ΔH_f°(kJ/mole)	S° (J/mole·K)
$SO_2(g)$	-296.8	248.2
$O_2(g)$	0	205.1
$SO_3(g)$	-395.7	256.8

PRACTICE PROBLEM SOLUTIONS

1. a. decrease, the solid is more ordered than the
      liquid.
   b. decrease, a less complex species is formed.
   c. increase, a solid dissolved in a liquid.
   d. decrease, a gas dissolving in a liquid.

2. Begin by writing the equation for the conversion
of diamond to graphite.

$$C(diamond) \rightarrow C(graphite)$$

Remembering that graphite is the standard state for
carbon, calculate the free energy change for this
process.

$$\Delta G^{\circ} = \Delta G^{\circ} (graphite) - \Delta G^{\circ} (diamond)$$

$$= 0 - (2.90 \text{ kJ/mole})$$

$$\Delta G^{\circ} = -2.90 \text{ kJ/mole}$$

   The negative free energy change value
indicates that diamond is spontaneously changing
into graphite, but everyday experience suggests
that diamonds seem to be quite permanent.
   This apparent contradiction is explained by
the fact that thermodynamics tells us nothing about
the <u>rate</u> of processes. Even though the diamond
does tend to change into graphite, the rate of the
transformation is extremely slow.

3. The equation that relates enthalpy of
vaporization and entropy change is

$$\Delta S^{\circ} = \frac{H_{vap}}{T} = \frac{30.0 \times 10^3 \text{ J/mol}}{350.0^{\circ}C}$$

$$\underline{\Delta S^{\circ} = 86 \text{ J/K} \cdot \text{mole}}$$

As would be expected, the entropy change is
positive for the conversion of liquid molecules
into gaseous molecules.

4. Substituting into the equation

$$\Delta G^\circ = \Sigma m \Delta G^\circ \text{ (products)} - \Sigma n \Delta G^\circ \text{ (reactants)}$$

$$\Delta G^\circ_f = 4\ \Delta G^\circ_f\ (H_2O(g)) - \Delta G^\circ_f\ (N_2H_4(\ell)) - 2\ \Delta G^\circ_f\ (H_2O_2(\ell))$$

$$= 4(-228.6 \text{ kJ/mole}) - 149.2 \text{ kJ/mole}$$
$$- 2(-120.4 \text{ kJ/mole})$$

$$\underline{\Delta G^\circ = -822.8 \text{ kJ}}$$

The large negative value agrees with the idea of using hydrazine as a rocket fuel, since it suggests that there is a great deal of energy that could be released to do work.

5. a. Determine the entropy change for this reaction.

$$\Delta S^\circ = \Sigma m S^\circ \text{ (products)} - \Sigma n S^\circ \text{ (reactants)}$$
$$\Delta S^\circ = 2\ S^\circ(NO(g)) - S^\circ(O_2(g)) - S^\circ(N_2(g))$$
$$= 2(210.7 \text{ J/mol·K}) - (205.0 \text{ J/mol·K}) - (191.5 \text{ J/mol·K})$$
$$= 24.9 \text{ J/K}$$

Since two moles of NO are formed in the reaction

$$\Delta H^\circ = 2(90.25 \text{ kJ/mol}) = 180.5 \text{ kJ}$$

Substituting these values into the equation

$$\Delta G^\circ = \Delta H^\circ - T\ \Delta S^\circ$$

$$= 180.5 \text{ kJ} - (298 \text{ K})(24.9 \text{ J/K})(1 \times 10^{-3} \text{ kJ/J})$$

$$\underline{\Delta G^\circ = 173.1 \text{ kJ}}$$

The positive value indicates that this reaction will not be spontaneous at 25°C.

b. As the temperature increases, the entropy term will have a greater tendency to make the reaction spontaneous. To find the temperature at which the reaction will be at equilibrium, set ΔG° equal to zero.

$$\Delta G^{\circ} = 0 = \Delta H^{\circ} - T \Delta S^{\circ}$$

Now solve the equation for T

$$T = \frac{\Delta H^{\circ}}{\Delta S^{\circ}} = \frac{180.5 \text{ kJ}}{0.0249 \text{ kJ/K}}$$

$$\underline{T = 7250 \text{ K}}$$

<u>The reaction will be spontaneous at temperatures above 7250 K.</u>

However, remember that even at lower temperatures the reaction may occur to a significant extent if there is little or no nitrogen oxide present initially.

6. The relationship between ΔG and K is given by the equation

$$\Delta G^{\circ} = - RT \ln K$$

Rearranging to solve for K

$$\ln K = \frac{-\Delta G}{RT} = \frac{-(-100.4 \text{ kJ/mol})(1 \times 10^3 \text{ J/kJ})}{(8.314 \text{ J/K} \cdot \text{mol})(298 \text{ K})}$$

$$\ln K = 40.5$$

$$\underline{K = 4.0 \times 10^{17}}$$

7. First, calculate the entropy change for the reaction.

$$\Delta S^{\circ} = 2 \text{ } S^{\circ}(SO_3(g)) - 2 \text{ } S^{\circ}(SO_2(g)) - S^{\circ}(O_2(g))$$

$$= 2(256.8 \text{ J/mol} \cdot \text{K}) - 2(248.1 \text{ J/mol} \cdot \text{K}) - (205.1 \text{ J/mol} \cdot \text{K})$$

$$\Delta S^{\circ} = -187.7 \text{ J/K}$$

Now, calculate the enthalpy change

$$\Delta H^{\circ} = 2 \text{ } \Delta H^{\circ}_f (SO_3(g)) - 2 \text{ } \Delta H^{\circ}_f (SO_2(g)) - \Delta H^{\circ}_f (O_2(g))$$

$$= 2(-395.7 \text{ kJ/mol}) - 2(-296.8 \text{ kJ/mol}) - (0)$$

$$\Delta H^{\circ} = -197.8 \text{ kJ}$$

Now, calculate the value of the free energy change

$$\Delta G^{\circ} = \Delta H^{\circ} - T \Delta S^{\circ}$$
$$= (-197.8 \times 10^{3} \text{ J}) - 298 \text{ K}(-187.7 \text{ J/K})$$
$$\Delta G^{\circ} = -141,800 \text{ J}$$

Finally, use the free energy value to determine K

$$\Delta G^{\circ} = - RT \ln K$$

$$-141,800 \text{ J} = -(8.314 \text{ J/K} \cdot \text{mol})(298 \text{ K}) \ln K$$

$$\ln K = 57.2$$

$$K = \underline{\quad 1.43 \times 10^{25} \quad}$$

PRACTICE TEST (40 Minutes)

1. Without looking up any numerical values, predict whether the entropy will be increase or decrease for each of the following reactions:

a. $Br_2(\ell) \qquad \rightarrow \qquad Br_2(g)$

b. $2 \ Cl(g) \qquad \rightarrow \qquad Cl_2(g)$

c. $Ag^{+}(aq) + Cl^{-}(aq) \rightarrow AgCl(s)$

d. $H_2O(\ell) \qquad \rightarrow \qquad H_2O(s)$

e. $O_2(g) \qquad \rightarrow \qquad O_2(aq)$

2. Both entropy and enthalpy make a significant contribution to determining whether or not a reaction will be spontaneous. Sometimes these two factors both have the same effect, but sometimes they work in opposite directions. In the latter case, temperature can be the deciding factor that determines reaction spontaneity. For each of the following cases, examine the sign of the enthalpy and entropy changes and suggest whether a process with those values would probably be spontaneous,

non-spontaneous, spontaneous at low temperatures, or spontaneous at high temperatures.

ΔH	ΔS	Prediction
+	+	
+	−	
−	+	
−	−	

3. The formation of photochemical smog is a complex process that involves hundreds of reactions. One of the important steps in this system is represented by the equation

$$O_3(g) \;+\; NO(g) \;\rightarrow\; NO_2(g) \;+\; O_2(g)$$

a. Calculate the value of ΔG° for this reaction using the thermodynamic values provided below.

b. Calculate the value of K for this reaction.

Standard Free Energies of Formation

$O_3(g)$ 163 kJ/mol $NO(g)$ 86.57 kJ/mol
$NO_2(g)$ 51.30 kJ/mol

4. Calculate the molar entropy of vaporization for liquid ammonia at its boiling point ($-33^\circ C$). The enthalpy of vaporization for ammonia at the boiling point is 23. kJ/mol.

5. Nitrous oxide, N_2O, is used as a propellant for soft ice cream and as an anesthetic. Calculate the standard free energy of formation for this reaction and determine if it seems reasonable for the commercial production of this gas?

$$2\,N_2O_4(g) \;\rightarrow\; 2\,N_2O(g) \;+\; 3\,O_2(g)$$

Standard Free Energies of Formation (kJ/mol)

$N_2O_4(\ell)$ 97.9 $N_2O(\ell)$ 104.2

6. At 25°C, the standard enthalpy of reaction is -91.3 kJ for the reaction

$$CCl_4(g) \ + \ H_2(g) \ \rightleftarrows \ HCl(g) \ + \ CHCl_3(g)$$

Using the absolute entropy values provided, determine whether or not this reaction is spontaneous at 25°C.

Absolute Entropy Values (J/mol·K)

$CCl_4(g)$ 309.7 $H_2(g)$ 130.6

$HCl(g)$ 186.8 $CHCl_3(g)$ 295.6

7. Using the standard free energies of formation provided below, calculate K at 25°C for the reaction

$$CO_2(g) \ + \ H_2(g) \ \rightleftarrows \ CO(g) \ + \ H_2O(g)$$

Standard Free Energies of Formation (kJ/mole)

$CO(g)$ -137.2; $H_2O(g)$ -228.6; $CO_2(g)$ -394.4

CONCEPT TEST ANSWERS

1. spontaneous
2. a. decrease in energy (or enthalpy)
   b. increase in disorder or chaos (i.e. entropy).
3. entropy
4. zero, third law of thermodynamics
5. A,b, and d represent an increase in entropy;
   c represents a decrease in entropy.
6. combined entropy   7. zero
8. negative   9. zero
10. 8.314 J/K·mol

PRACTICE TEST ANSWERS

1. (L1) a. increase, liquid converted to a gas
   b. increase, a more complex species forms
   c. decrease, a precipitate forms
   d. decrease, liquid changes to a solid
   e. decrease, gas dissolves in a solvent

2. (**L3**)

ΔH	ΔS	Prediction
+	+	spontaneous at high temperatures
+	−	non-spontaneous
−	+	spontaneous
−	−	spontaneous at low temperatures

3. a. (**L2**) −198.0 kJ
   b. (**L4**) $K = 5.6 \times 10^{34}$

4. (**L1**) 96 J/K·mole

5. (**L2**) $\Delta G^{\circ} = 12.6$ kJ. The positive free energy for this reaction suggests that it would not be useful, at least under these conditions.

6. (**L3**) $\Delta S^{\circ} = 42.0$ J/K; $\Delta G^{\circ} = -105.0$ kJ; The reaction is spontaneous

7. (**L4**) $\Delta G = 28.6$ kJ; $K = 1.03 \times 10^{-5}$

CHAPTER 21
ELECTROCHEMISTRY: THE CHEMISTRY OF OXIDATION-REDUCTION REACTIONS

LEARNING GOALS:

1. Understand how an oxidation-reduction reaction in a voltaic cell can be used to produce an electric current, recognize the various components of such a cell (anode, cathode, and salt bridge). (Sec. 21.1)

2. Since ΔG is negative for spontaneous reactions, the equation

$$\Delta G = -nFE^{\circ}$$

indicates that all spontaneous electron transfer reactions have a positive value for E°. Be able to use a table of standard reduction potentials (such as Table 21.1 in the textbook) to calculate the E° value for the combination of a pair of half-reactions and use this result to predict whether or not this combination will occur spontaneously under standard conditions. (Sec. 21.2)

3. When the conditions in an electrochemical cell are nonstandard, be able to use the Nernst equation,

$$E = E^{\circ} - \frac{2.303RT}{nF} \log Q$$

to calculate the potential of an electrochemical cell. (Sec. 21.3)

4. Also be able to use the Nernst equation to determine equilibrium constants from standard reduction potential values. (Sec. 21.3)

5. A variety of different voltaic cells are commonly used to provide electric current, including the dry cell, the storage battery, the mercury battery, the alkaline battery, and fuel cells. Be familiar with the chemistry involved in some of these systems. (Sec. 21.4)

6. Electrolysis is an important commercial technique, and so it is important to be able to predict the products that result when an electric current is passed through various types of solutions or melted solids. Be able to use standard reduction potentials to predict the most likely products of electrolysis reactions and also be familiar with some of the important commercial applications of electrolysis. (Sec. 21.5)

7. The Faraday constant serves as the basis for a useful relationship between current flow and the amount of chemical reaction that can occur. It's important to be able to do these conversions in either direction and also to be able to determine the amount of energy involved in units of either joules or kilowatt-hours. (Sec. 21.6)

8. Corrosion reactions are an extremely important type of oxidation-reduction process. Understand the conditions that are most likely to produce corrosion as well as how corrosion can be prevented. (Sec. 21.8)

IMPORTANT NEW TERMS:

alkaline battery (21.4)
ampere (21.6)
anode (21.1)
anodic inhibition (21.8)
battery (21.4)
cathode (21.1)
cathodic protection (21.8)
cell voltage (or cell potential) (21.2)
corrosion (21.8)
coulomb (21.6)
current (21.6)
dry cell (21.4)
electrode (21.1)
electrolysis (21.5)
electromotive force (or emf) (21.1)
Faraday constant (21.6)
fuel cell (21.4)
lead storage battery (21.4)
mercury battery (21.4)
Nernst equation (21.3)
NI-CAD battery (21.4)

348

overvoltage (21.5)
sacrificial anode (21.8)
salt bridge (21.1)
standard hydrogen electrode (S.H.E.) (21.2)
standard reduction potential (21.2)

CONCEPT TEST

1. In an electrochemical cell, oxidation always occurs at the electrode called the _____, and reduction always occurs at the electrode called the _____.

2. In order to maintain a balance of ion charges in the compartments of an electrochemical cell, the two half-cells are often connected with a device called a(n) _____.

3. The value of E^o for all spontaneous electron transfer reactions must be _____.

4. Using Table 21.1, arrange the following ions in order of increasing strength as oxidizing agents: Ag^+, Na^+, and Sn^{2+}.

(weakest) _____ < _____ < _____ (strongest)

5. Unless specified otherwise, all values of E^o are given at a temperature of _____.

6. The standard cell potentials for all half-cell reactions is measured relative to the _____ _____ electrode.

7. Determine the value of n (as used in the Nernst equation) for each equation. (Be sure the equation is balanced first!)

a. Cu^{2+} + Cd $\rightarrow$ Cd^{2+} + Cu

b. Al + Sn^{2+} $\rightarrow$ Al^{3+} + Sn

c. Na + Cl_2 $\rightarrow$ Cl^- + Na^+

8. When the cell potential, E, has a value of _____, the Q term in the Nernst equation is equivalent to the equilibrium constant, K.

9. Sodium metal is produced in a Downs cell by the electrolysis of _____.

10. Chlorine is produced by the electrolysis of _____.

11. Protecting a metal object from corrosion by painting the surface or forming an oxide coating is called _____.

12. If a substance has a reduction potential more negative than _____ volts, then only water will be reduced when we pass an electric current through an aqueous solution of this species.

13. In a battery, the polarity of the anode is ___, but for an electrolysis cell, the polarity of the anode is ___. (Choose from + or - in each case.)

14. In many cases, the current measured when a cell is working is found to be different from that calculated, even though the cell componenets are at standard conditions. This difference is called _____, and is primarily due to the effect of reaction kinetics.

STUDY HINTS:

1. Remember that in the Nernst equation, the value of n is determined by the number of electrons transferred when balancing the net equation. Normally is isn't possible to simply look at one half reaction and determine what the n value will be for a net equation. Be sure to balance the two half reactions in the usual way, and then determine n from the number of electrons that are canceled out when the two half reactions are added.

2. In many cases, chemists have agreed to always do certain things the same way when writing electrochemistry problems. For example, the anode is normally written on the left on an electrochemical cell, and the table of standard electrode potentials is normally written with the most negative potentials on the top of the list. Don't depend too much on these conventions. It can be a rude shock for those who memorize that the strongest oxidizing agents are at the top of the reduction potential table, then encounter a table that lists the half reactions in the opposite order. It's always better to try to understand rather than just memorize isolated facts, but that is especially true in electrochemistry.

3. When using the table of standard reduction potentials, don't forget that the half-reactions listed include both an oxidizing agent and a reducing agent. When asked to identify the oxidizing agent in a process, if you give the complete half-reaction, the answer contains both an oxidizing agent and a reducing agent and may be marked wrong for that reason.

PRACTICE PROBLEMS

1. (**L1**) Using Table 21.1 in your text, draw a simple diagram of each voltaic cell below. Assume the metal ions are in 1 Molar aqueous solution and use an inert platinum electrode if necessary. Label the cells thoroughly, including the anode and cathode, the direction that the electrons move in the circuit, and assuming that the salt bridge contains $NaNO_3$, the direction the NO_3^- ions move.

A. $Cd(s) \mid Cd^{2+}(1\ M) \parallel Sn^{2+}(1\ M) \mid Sn$

B. $Fe(s) \mid Fe^{2+}(1\ M) \parallel Cu^{2+}(1\ M) \mid Cu$

2. **(L5)** Briefly describe the following common batteries, including identifying the anode and cathode. (a) dry cell (b) mercury cell

3. **(L2)** Use the following table of reduction potentials to answer the various questions below.

Half Reaction	$E^o(V)$
$Cl_2(g) + 2\ e^- \rightarrow 2\ Cl^-(aq)$	+1.360
$Ag^+(aq) + e^- \rightarrow Ag(s)$	+0.80
$Cu^{2+}(aq) + 2\ e^- \rightarrow Cu(s)$	+0.337
$Ni^{2+}(aq) + 2\ e^- \rightarrow Ni(s)$	-0.25
$Cd^{2+}(aq) + 2\ e^- \rightarrow Cd(s)$	-0.40
$Mg^{2+}(aq) + 2\ e^- \rightarrow Mg(s)$	-2.37

Based only on the above table:

a. _____ is the strongest oxidizing agent in the above table.
b. _____ is the strongest reducing agent in the above table.
c. _____ are the symbols of the metals from the table that will reduce copper(II) ion.
d. _____ are the symbols of metals from the table can be oxidized by silver(I) ion.
e. Will chlorine gas oxidize magnesium metal to magnesium(II) ion? _____

4. **(L7)** A current of 2.50 amperes was passed through a solution of chromium(III) chloride for 3.50 hours. What mass of chromium metal was deposited on the cathode of the electrolytic cell?

5. **(L3)** Using the table of reduction potentials in your text, calculate the voltage for the following electrochemical cell:

$$Cd(s) \mid Cd^{2+}(0.20 \text{ M}) \parallel Sn^{2+}(0.80 \text{ M}) \mid Sn(s)$$

6. **(L4)** The standard reduction potential equals −1.245 volts in basic solution for the half reaction

$$Zn(OH)_2(s) + 2 e^- \rightarrow Zn(s) + 2 OH^-(aq)$$

Using Table 21.1 in the text, select an appropriate half reaction and determine the K_{sp} value for $Zn(OH)_2$.

7. (**L6**) As described in your textbook, aluminum is prepared by the Hall-Heroult process, which involves the electrolysis of a molten mixture of aluminum oxide and cryolite, Na_3AlF_6. a. How many coulombs of electricity would be required to produce 1000 kilograms of aluminum metal? b. If the electrolysis cell operates at 5.0 volts, how many kilowatt-hours of electricity will be required to produce this much aluminum?

PRACTICE PROBLEM SOLUTIONS

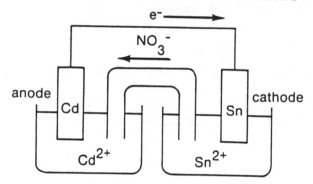

Figure 21.1 Answer to Practice Problem 1a.

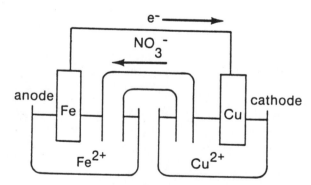

Figure 21.2 Answer to Practice Problem 1b.

2. The common dry cell contains a carbon rod electrode inserted into a moist paste of NH_4Cl, $ZnCl_2$, and MnO_2 in a zinc can that serves as the anode.

The mercury battery, which is used in many devices that require a small cell, uses metallic zinc as the anode and mercury (II) oxide as the cathode.

3. a. $Cl_2(g)$ is the strongest oxidizing agent
b. $Mg(s)$ is the strongest reducing agent
c. Magnesium, cadmium, and nickel will reduce copper(II).
d. Copper, nickel, cadmium, and magnesium are oxidized by silver(I).
e. The E^o value for the reaction would be +3.73 volts, so the reaction is very spontaneous.

4. Since you are asked to find the grams of chromium, the calculations should be planned to obtain the moles of chromium. This can be calculated if you know the moles of electrons that have been transferred. That will depend on the current and the time. This sequence of calculations can be represented as follows:

Time x current $\rightarrow$ charge $\rightarrow$ moles $\rightarrow$ moles
of electrons of Chromium

First, determine the number of coulombs that passed through the solution in 3.50 hours.

Charge (coulombs) = amps x time(sec)

= 2.50 amps x 3.50 hr x 60.0 min/hr x 60 s/min

= 31,500 coulombs

The half reaction for deposition of the chromium is

$$Cr^{3+}(aq) + 3e \rightarrow Cr(s)$$

Therefore the deposition requires three moles of electrons per mole of chromium deposited.

The number of moles of electrons is then

355

$$\text{Faradays} = x \ \frac{(31{,}500 \text{ coulombs})}{(96{,}500 \text{ coulombs/Faraday})} \ x \ \frac{1 \text{ mole Cr}}{3 \text{ Faradays}}$$

Moles of Chromium = 0.109 moles

Now determine the mass of chromium produced.

Mass of Cr = 0.109 moles x 52.00 g/mole

<u>Mass of Cr = 5.66 grams</u>

5. From Table 21.1, the voltages of the half reactions are:

$$Cd^{2+} + 2e^- \rightarrow Cd(s) \qquad E^o = -0.40 \text{ Volts}$$

$$Sn^{2+} + 2e^- \rightarrow Sn(s) \qquad = -0.14$$

Reverse the cadmium half reaction, and add the half reactions to produce the overall reaction:
Oxidation

$$Cd(s) \rightarrow Cd^{2+} + 2e^- \qquad E^o = +0.40 \text{ V}$$

Reduction

$$\underline{Sn^{2+} + 2e^- \rightarrow Sn(s) \qquad E^o = -0.14 \text{ V}}$$

$$Sn^{2+} + Cd(s) \rightarrow Cd^{2+} + Sn(s) \quad E^o_{cell} = +0.26 \text{ V}$$

Since the concentrations of the reactants are not 1 M, it is necessary to use the Nernst equation to determine the effect of the nonstandard concentrations.

$$E = E^o - \frac{2.303RT}{nF} \log Q$$

$$E = E^o - \frac{0.0592}{n} \log \frac{[Cd^{2+}]}{[Sn^{2+}]}$$

$$= 0.26 \text{ V} - \frac{0.0592}{2 \text{ mol electrons}} \log \frac{0.20M}{0.80M}$$

E = 0.26 V + 0.0178 V

$E_{cell} = \underline{0.28\ V}$

The result of the concentration change was to make the cell reaction more spontaneous.

6. If you combine the half reaction provided with the half reaction for zinc ion, you will obtain the solubility product reaction for $Zn(OH)_2$:

$$Zn(OH)_2(s) + 2\ e^- \rightarrow Zn(s) + 2\ OH^-(aq) \quad E^o = -1.245\ V$$

$$Zn(s) \quad \rightarrow \quad Zn^{2+} + 2\ e^- \quad\quad\quad\quad = \quad 0.763\ V$$

$$Zn(OH)_2(s) \rightarrow \quad Zn^{2+}(aq) \quad + \quad 2\ OH^-(aq) \quad E^o = -0.482\ V$$

Once you have determined the value of E^o for the reaction, it is easy to substitute into the Nernst equation and solve for K:

$$E = E^o - \frac{2.303RT}{nF} \log K$$

Setting E = 0 at equilibrium

$$E^o = \frac{2.303RT}{nF} \log K$$

$$\log K = \frac{(2.0\ mole)(-0.482\ V)}{0.0592} = \frac{-\ 0.964\ V \cdot mole}{0.0592\ V \cdot mole}$$

$$\log K = -\ 16.28$$

$$\underline{K = 5.2 \times 10^{-17}}$$

7. First, calculate the moles of aluminum to be formed.

$$Moles\ of\ aluminum = \frac{1000\ kg \times 1 \times 10^3\ g/kg}{27.0\ g/mole}$$

$$Moles\ of\ aluminum = 3.7 \times 10^4\ moles$$

357

The half reaction for the deposition of aluminum is

$$Al^{3+}(melt) + 3e^- \rightarrow Al(s)$$

3 moles of electrons will produce 1 mole of Al.

$$mole\ of\ e^- = \frac{(3.7 \times 10^4\ moles\ of\ Al)(3\ moles\ of\ e^-)}{(1\ mole\ Al)}$$

moles of e^- required = 1.11×10^5 moles of e^-

Next determine the coulombs needed

$$coulombs = (1.11 \times 10^5\ moles\ of\ e^-)\frac{(96,500\ coulombs)}{mole\ of\ electrons}$$

$\underline{coulombs = 1.1 \times 10^{10}\ coulombs}$

6. b. To calculate the number of kilowatt-hours, first determine the number of joules

Joules = coulombs x volts

= $(1.1 \times 10^{10}\ coulombs)(5.0\ V)$

Joules = 5.4×10^{10} joules

Now calculate the number of kilowatt-hours.

kwh = $(5.4 \times 10^{10}\ joules)(1\ kwh/3.6 \times 10^6\ J)$

$\underline{kilowatt\text{-}hours = 1.5 \times 10^4\ kwh}$

PRACTICE TEST (45 min.)

1. Using Table 21.1 in your text, draw a simple diagram of the voltaic cell given below. Use an inert platinum electrode if necessary and assume that the dissolved ions are in 1 Molar aqueous solution. Label the cell thoroughly, including the anode and cathode, the direction of motion of the electrons in the circuit, and assuming that the salt bridge contains $NaNO_3$, the direction of movement of the NO_3^- ions.

A. $Al(s) + Zn^{2+} \rightarrow Al^{3+} + Zn(s)$

2. First, balance each equation below and then use Table 21.1 in your textbook to calculate the standard potential, E°, and predict which of these reactions will occur spontaneously in the direction in which they are written.

a. $Al(s) + Cu^{2+}(aq) \rightarrow Al^{3+}(aq) + Cu(s)$

b. $Zn(s) + Mg^{2+}(aq) \rightarrow Zn^{2+}(aq) + Mg(s)$

c. $Cd(s) + Ag^+(aq) \rightarrow Cd^{2+}(aq) + Ag(s)$

d. $Hg^{2+}(aq) + PbSO_4(s) + H_2O(\ell)$
$\rightarrow Hg(\ell) + PbO_2(s) + SO_4^{2-}(aq) + H^+(aq)$

3. A current of 2.00 amps is passed through a solution of copper nitrate until 6.35 grams of copper metal has been deposited. a. How many seconds did the current flow? b. How many hours did the current flow?

4. An electrochemical concentration cell is set up as shown:

$$Cu(s) \mid Cu^{2+}(0.10\ M) \parallel Cu^{2+}(10.0\ M) \mid Cu(s)$$

What is the initial voltage of this cell when it is operated with the concentrations indicated?

5. Determine the cell voltage for the following cell

$$Zn(s) \mid Zn^{2+}(0.10\ M) \parallel Ag^+(0.0010\ M) \mid Ag(s)$$

if the voltage for this cell under standard conditions, E°, is 1.56 volts.

6. The standard reduction potential equals 0.222 volts for the half reaction

$$AgCl(s) + e^- \rightarrow Ag(s) + Cl^-(aq)$$

Using Table 21.1 in the text, select an appropriate half reaction and determine the K_{sp} value for AgCl.

7. Indicate the chemical reaction at the anode and

359

the cathode if an electric current is passed through each of the following:

a. $MgCl_2$(aqueous)
c. LiBr(aqueous)
b. $MgCl_2$(molten)
d. $CdCl_2$(aqueous)

CONCEPT TEST ANSWERS

1. anode, cathode
2. salt bridge
3. a positive number
4. (weakest) Na^+ < Sn^{2+} < Ag^+ (strongest)
5. $25°C$
6. standard hydrogen electrode (or S.H.E.)
7. a. two   b. six   c. two
8. zero
9. molten sodium chloride
10. aqueous sodium chloride solution
11. anodic inhibition
12. about -0.8 volts
13. negative (-), positive (+)
14. overvoltage

PRACTICE TEST ANSWERS

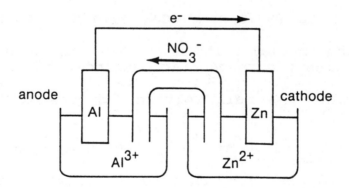

Figure 21.2 Answer to practice test, question 1.
2. (L2)
a. 2 Al(s) + 3 Cu^{2+}(aq) → 2 Al^{3+}(aq) + 3 Cu(s)
   E^o = +1.997 V, spontaneous

b. Mg^{2+}(aq) + Zn(s)   → Mg(s) + Zn^{2+}(aq)
   E^o = -1.607 V, nonspontaneous

c. Cd(s) + 2 Ag^+(aq) → Cd^{2+}(aq)+ 2 Ag(s)
   E^o = +1.20 V, spontaneous

360

d. Hg^{2+}(aq) + $PbSO_4$(s) + 2 H_2O(ℓ)
        $\rightarrow$ Hg(ℓ) + PbO_2(s) + SO_4^{2-} + 4 H^+
   E° = -0.830 V, nonspontaneous

3. **(L7)** a. 9640 seconds, b. 2.68 hours
4. **(L3)** 0.0592 V
5. **(L3)** voltage = 1.42 volts
6. **(L4)** 1.7×10^{-10}
7. a. Anode (oxidation of chloride)

   2 Cl^-(aq) $\rightarrow$ Cl_2(g) + 2 e^-
   Cathode (reduction of water)

   2 H_2O(ℓ) + 2 e^- $\rightarrow$ H_2(g) + 2 OH^-(aq)
b. Anode (oxidation of chloride)

   2 Cl^-(aq) $\rightarrow$ Cl_2(g) + 2 e^-
   Cathode (reduction of magnesium)

   Mg^{2+}(aq) + 2 e^- $\rightarrow$ Mg(s)
c. Anode (oxidation of bromide)

   2 Br^-(aq) $\rightarrow$ Br_2(g) + 2 e^-
   Cathode (reduction of water)

   2 H_2O(ℓ) + 2 e^- $\rightarrow$ H_2(g) + 2 OH^-(aq)
d. Anode (oxidation of chloride)

   2 Cl^-(aq) $\rightarrow$ Cl_2(g) + 2 e^-
   Cathode (reduction of cadmium)

   Cd^{2+}(aq) + 2 e^- $\rightarrow$ Cd(s)

CHAPTER 22
THE CHEMISTRY OF HYDROGEN AND
THE s-BLOCK ELEMENTS

LEARNING GOALS:

1. Learn about the chemical and physical properties, methods of preparation, compounds, and uses of hydrogen. (Sec. 22.1

2. The diagonal relationship is an important addition to the periodic relationships studied earlier. Watch for examples of this type of behavior in this and the next chapter and understand why this behavior exists. (Sec. 22.1)

3. Be familiar with the properties, methods of preparation, common compounds, and uses of the alkali metals. (Sec. 22.3)

4. Also know the properties, methods of preparation, common compounds, and uses of the alkaline earth metals. (Sec. 22.4)

IMPORTANT NEW TERMS:

anionic hydride (22.1)
baking soda (22.3)
chlor-alkali industry (22.3)
covalent hydride (22.1)
deuterium (22.1)
diagonal relationship (22.2)
interstitial metallic hydride (22.1)
peroxide (22.3)
protium (22.1)
slaked lime (22.3)
soda ash (22.3)
Solvay process (22.3)
steam reformation of hydrocarbons (22.1)
superoxide (22.3)
tritium (22.1)
washing soda (22.3)
water gas or synthesis gas (22.1)

CONCEPT TEST

1. Hydrogen has three isotopes, which are called
_____, _____, and _____.

2. Hydrogen will form chemical combinations with
every element except those in the periodic group
called the _____.

3. The largest commercial use of hydrogen gas is
the Haber process, which is a major source of
_____ gas.

4. _____ is the only alkali metal
that forms a simple oxide of the formula M_2O when
reacted with oxygen.

5. The process called _____ consists of
adding hydrogen to vegetable oil in order to
convert carbon-carbon double bonds into single
bonds. Commercial products produced in this way
include peanut butter, margarine, and cooking fats.

6. According to the diagonal relationship, the
chemistry of boron should have some similarities to
the chemistry of the element _____.

7. Among the alkali metals, _____ is the
best reducing agent, and _____ is the
poorest reducing agent.

8. Match each chemical formula on the left with its
common name on the right.

_____ $NaHCO_3$ A. washing soda
_____ $Ca(OH)_2$ B. soda ash
_____ Na_2CO_3 C. baking soda
_____ $Na_2CO_3 \cdot 10\ H_2O$ D. slaked lime

9. The mineral beryl is colorless when pure, but if
some of the aluminum(III) ions are replaced by

chromium(III) ions, it has a beautiful green color and is called a(n) _____.

10. The principal source of commercial calcium compounds is the mineral _____.

11. There are several steps in the commercial preparation of magnesium metal. First, a base, such as calcium hydroxide, is added to sea water and the magnesium precipitates in the form of the compound _____. After filtration, hydrochloric acid is added, and the water evaporated to leave _____, which can then be electrolyzed to form the metal.

12. The chemistry of the alkali metals is dominated by the _____ oxidation state; the chemistry of the alkaline earth metals by the _____ oxidation state.

13. Water that contains significant amounts of the ions _____ and _____ is called hard water.

14. When lime is heated with coke, the product is _____, which has the formula _____.

15. Which alkaline earth metal is a component of the chlorophyll molecule? _____

STUDY HINTS:

1. It will be easier to remember the equations in the next few chapters if general types of reactions are identified and learned together rather than just memorizing each equation separately. For instance, try to pick out all the examples of the following cases in this chapter:
   (a) direct combination of one element with another,
   (b) reduction with a reducing agent, such as carbon monoxide or hydrogen, and
   (c) replacement of a less active metal with a more active metal.

PRACTICE PROBLEMS

1. (**L1, L3, & L4**) Complete and balance the following equations:

a. $H_2(g)$ + $Cl_2(g)$ $\rightarrow$

b. $H_2O(g)$ + $CO(g)$ $\overset{heat}{\rightarrow}$

c. $Na(s)$ + H_2 $\rightarrow$

d. $Na(g)$ + $KCl(\ell)$ $\rightarrow$

e. $Li(s)$ + $O_2(g)$ $\rightarrow$

f. $Na(s)$ + $O_2(g)$ $\rightarrow$

g. $Ba(s)$ + $O_2(air)$ $\rightarrow$

h. $CaC_2(s)$ + $H_2O(\ell)$ $\rightarrow$

i. $NH_4Cl(aq)$ + $CaO(s)$ $\rightarrow$

2. (**L4**) Arrange the following ions in order of decreasing ionic radius: Mg^{2+}, Be^{2+}, Ba^{2+}, and Sr^{2+}.

3. (**L1**) Arrange the following compounds in order of increasing extent of hydrogen bonding (as evidenced by the size of the deviation from the expected boiling point): NH_3, HF, CH_4, and CF_3H. (If necessary, review the discussion of hydrogen bonding in Section 13.2 of the text.)

4. (**L1**) List at least three elements that will liberate hydrogen gas from dilute acidic solution. (Hint: Use the Table of reduction potentials in Chapter 21.)

5. **(L1 & L3)** Indicate whether each of the following species is most likely to act as a Lewis Acid or a Lewis Base.

a. Na^+ _____   b. PH_3 _____

c. NH_3 _____   d. $C_2H_5^-$ _____

6. **(L1, L3, & L4)** Suggest a method for the preparation of each of the following compounds by means of a simple, exchange reaction. (Hint: If necessary, review the solubility rules in Chapter 5.)

a. $BaCO_3$ b. $NaNO_3$

c. $SrSO_4$ d. $Ca(ClO_3)_2$

a.

b.

c.

d.

7. **(L1, L3, & L4)** Write balanced chemical reactions for one method of preparation for each substance listed.

a. ammonia

b. overall reaction for the Solvay process forming soda ash

c. preparation of tetraethyl lead

PRACTICE PROBLEM SOLUTIONS

1. a. $H_2(g)$ + $Cl_2(g)$ → $2 HCl(g)$

b. $H_2O(g)$ + $CO(g)$ $\xrightarrow{heat}$ $H_2(g)$ + $CO_2(g)$
Depending on conditions, CH_3OH may be formed.

c. $2 Na(s)$ + $H_2(g)$ → $2 NaH(s)$

d. $Na(g)$ + $KCl(\ell)$ → $K(g)$ + $NaCl(\ell)$

e. $4 Li(s)$ + $O_2(g)$ → $2 Li_2O(s)$

f. $2 Na(s)$ + $O_2(g)$ → $Na_2O_2(s)$

g. $2 Ba(s)$ + $O_2(air)$ → $2 BaO(s)$

h. $CaC_2(s)$ + $2 H_2O(\ell)$ → $Ca(OH)_2(aq)$ + $H-C{\equiv}C-H(g)$

i. $2 NH_4Cl(aq)$ + $CaO(s)$
→ $2 NH_3(g)$ + $H_2O(\ell)$ + $CaCl_2(aq)$

2. Ba^{2+} > Sr^{2+} > Mg^{2+} > Be^{2+}

3. CH_4 < CF_3H < NH_3 < HF (Based on the increasing electronegativity of the element or group of elements attached to the hydrogen.)

4. Consult the table of reduction potentials found in Chapter 21. Those metal half reactions that are below the hydrogen half reaction would be good choices. Examples of such metals might be aluminum, zinc, calcium, sodium, and potassium. (Notice that the last two would be extremely reactive.)

5. a. Na^+ has no electron pairs that are readily available for donation but does have empty orbitals that can accept electron pairs. It is most likely to be a <u>Lewis Acid</u>.

b. PH_3 has an electron lone pair, but it also has empty d orbitals available that can accept electrons. It can be <u>either a Lewis Acid or a Lewis Base.</u>

c. NH_3 has one electron lone pair that makes it a
Lewis Base.

d. $C_2H_5^-$ has one electron lone pair that makes it a
Lewis Base.

6. In each case there are several acceptable
reactions that can be used to prepare the compound.
Be sure that both of the reactants are soluble, and
that at least one of the products is insoluble.

a. $BaCl_2(aq) + Na_2CO_3(aq) \rightarrow BaCO_3(s) + 2 NaCl(aq)$

b. $NaCl(aq) + AgNO_3(aq) \rightarrow NaNO_3(aq) + AgCl(s)$

c. $Sr(NO_3)_2(aq) + K_2SO_4(aq) \rightarrow SrSO_4(s) + 2 KNO_3(aq)$

d. $CaBr_2(aq) + Pb(ClO_3)_2(aq) \rightarrow Ca(ClO_3)_2(aq) + PbBr_2(s)$

7. a. $N_2(g) + 3 H_2(g) \rightarrow 2 NH_3(g)$

b. $2 NaCl + CaCO_3 \rightarrow Na_2CO_3 + CaCl_2$

c. $4 Na(s) + 4 C_2H_5Cl(\ell) + Pb(s)$
$\rightarrow 4 NaCl(s) + Pb(C_2H_5)_4(\ell)$

PRACTICE TEST (30 Minutes)

1. Arrange the following metallic elements in order
of their increasing reactivity with water:
beryllium, sodium, cesium, and magnesium.

2. Complete and balance the following equations:

a. $C(s) + H_2O(g) \xrightarrow{heat}$

b. $K(s) + H_2O(\ell) \rightarrow$

c. $TiCl_4(\ell) + Na(s) \rightarrow$

d. $K(s) + O_2(g) \rightarrow$

e. $CaF_2(s) + H_2SO_4(\ell) \rightarrow$

f. $CaO(s)$ + $C(s)$ $\xrightarrow{200^\circ C}$

g. $H_2(g)$ + $Ca(s)$ $\rightarrow$

h. $Zn(s)$ + $HCl(aq)$ $\rightarrow$

3. Arrange the following ions in order of increasing ionic radius: Cs^+, K^+, Li^+, and Na^+.

4. According to the text, hydrogen forms three different types of compounds. Name these three different types of compounds, and explain what type of elements are most likely to react with hydrogen and form these types of hydrides.

5. Indicate whether each of the following species is most likely to act as a Lewis Acid or a Lewis Base.

a. Be^{2+} b. BCl_3 c. NH_2^- d. SiH_4

6. Which of the species below is the Bronsted Acid of HPO_4^{2-}?

a. H_3PO_4 b. H_3O^+ c. $H_2PO_4^-$ d. PO_4^{3-}

7. Write balanced chemical reactions for one method of preparation for each substance listed.

a. methanol, CH_3OH

b. preparation of calcium metal by chemical reduction

c. production of beryllium from beryllium chloride

d. electrolysis of brine to form chlorine and sodium hydroxide

CONCEPT TEST ANSWERS

1. protium, deuterium, and tritium 2. rare gases
3. ammonia   4. lithium

369

5. hydrogenation   6. silicon
7. lithium is best, and sodium poorest
8. $NaHCO_3$ -baking soda $Ca(OH)_2$ - slaked lime,

   Na_2CO_3 - soda ash $Na_2CO3 \cdot 10\ H_2O$ -washing soda

9. emerald   10. limestone (or calcite)
11. magnesium hydroxide, $Mg(OH)_2$;
    magnesium chloride, $MgCl_2$
12. +1,+2

13. Ca^{2+} and Mg^{2+} 14. calcium carbide, CaC_2
15. magnesium

PRACTICE TEST ANSWERS

1. (**L3 & L4**) The order of reactivity is
beryllium < magnesium < sodium < cesium

2. (**L1, L3, & L4**)

a. $C(s)\ +\ H_2O(g)\ \xrightarrow{heat}\ H_2(g)\ +\ CO(g)$

b. $2\ K(s) + 2\ H_2O(\ell)\ \rightarrow 2\ KOH(aq)\ +\ H_2(g)$

c. $TiCl_4(\ell) + 4\ Na(s)\ \rightarrow\ Ti(s)\ + 4\ NaCl(s)$

d. $K(s)\ +\ O_2(g)\ \rightarrow\ KO_2(s)$

e. $CaF_2(s)+\ H_2SO_4(\ell) \rightarrow\ 2\ HF(g)\ +\ CaSO_4(s)$

f. $CaO(s) +\ 3\ C(s)\ \xrightarrow{200°C}\ CaC_2(s)\ +\ CO(g)$

g. $H_2(g)\ +\ Ca(s)\ \rightarrow\ CaH_2(s)$

h. $Zn(s)\ + 2\ HCl(aq)\ \rightarrow\ ZnCl_2(aq)\ +\ H_2(g)$

3. (**L3**) $Li^+ < Na^+ < K^+ < Cs^+$

4. (**L1**) The three types are (a) anionic hydrides,
formed when hydrogen reacts with metals of low
electronegativity, (b) covalent hydrides, formed
with electronegative elements, and (c) interstitial
metallic hydrides, when the small hydrogen atoms
are absorbed into the holes in the crystal lattice

370

of a metal.

5. (**L1, L3, & L4**) a. Be^{2+} has no electron pairs that are readily available for donation but does have empty orbitals that can accept electron pairs. It is most likely to be a <u>Lewis Acid</u>.

b. BCl_3 is an electron deficient molecule and may accept an electron pair, making it a <u>Lewis Acid</u>.

c. NH_2^- has two lone pairs that make it a <u>Lewis Base</u>.

d. SiH_4 has empty d orbitals that can accept electron pairs, and so it is a <u>Lewis Acid</u>.

6. (**L1**) c. $H_2PO_4^-$

7. (**L1, L3, & L4**)
$$\text{a. } 2\ H_2(g)\ +\ CO(g)\ \xrightarrow{\text{catalyst}}\ CH_3OH(\ell)$$

$$\text{b. } 3\ CaO(s)\ +\ 2\ Al(s)\ \rightarrow\ Al_2O_3(s)\ +\ 3\ Ca(s)$$

$$\text{c. } BeCl_2\ (\text{melt with } NaCl)\ \xrightarrow{\text{electrolysis}}\ Be(s)\ +\ Cl_2(g)$$

$$\text{d. } 2\ NaCl(aq)\ +\ 2\ H_2O(\ell)$$
$$\xrightarrow{\text{electrolysis}}\ 2\ NaOH(aq)\ +\ H_2(g)\ +\ Cl_2(g)$$

CHAPTER 23
METALS, METALLOIDS, AND NONMETALS:
PERIODIC GROUPS 3A AND 4A

LEARNING GOALS:

1. Be familiar with the properties, methods of preparation, common compounds, and uses of the group 3A elements. (Sec. 23.1)

2. Be familiar with the properties, methods of preparation, common compounds, and uses of the group 4A elements. (Sec. 23.2)

IMPORTANT NEW TERMS:

allotrope (23.1)
aluminosilicates (23.2)
Bayer process (23.1)
catenation (23.2)
chlorofluorocarbon (23.2)
clay mineral (23.2)
diborane (23.1)
dry ice (23.2)
Mond process 23.2)
silicate minerals (23.2)
three-center bond (23.1)
zeolite (23.2)
zone refining (23.1)

CONCEPT TEST

1. The element boron may exist in several different solid state structures called _____.

2. _____ is the only Group 3A element that is very hard, refractory, and a nonconductor.

3. Aluminum is highly corrosion resistant because a thin film of _____ forms rapidly on the surface of aluminum metal.

4. _____ is the element that has a melting point lower than human body temperature, a high boiling point, and is one of the few substances that expands upon freezing.

5. The mineral _____ is the most important boron-oxygen compound and also the most common form of boron in nature.

6. Although aluminum is widely used as a structural material and for packaging, the pure element is rarely used. Why? _____

7. In both Groups 3A and 4A, the lower oxidation numbers are found to be _____ (choose from more or less) stable for the heavier elements in the group.

8. Most of the lead oxide produced in this country each year is used to make _____ or

_____.

9. _____ are a special type of aluminosilicate having tunnels and cavities that make them useful a drying agents or catalysts.

10. _____ is the second most abundant element in the earth's crust.

11. Bronze is an alloy consisting primarily of the metals _____ and _____.

12. Silicon carbide, commonly called _____, is used extensively as an inexpensive abrasive.

13. Carbon is a good example of _____, the ability to form compounds based on the bonding together chains of atoms of the same element.

14. The simplest boron hydride has the formula _____ and is called _____.

15. The largest single use of carbon dioxide in this country is as a(n) _____, commonly known as _____.

STUDY HINTS:

1. As was the case in the previous chapter, it is much easier to memorize the many different reactions that are described if, as far as possible, they are systematically organized in terms of the types of reagents involved.

Reducing agents are especially important in this chapter, and so any reaction having a reactant such as carbon, carbon monoxide, or hydrogen, may well be most understandable in terms of a reduction of the other reactant.

Another common reaction type in this chapter involves direct combination of elements. The products in these cases should also usually be easy to predict and so not require much memorization.

Of course, not all of the reactions will fit these types, but try to identify those that do.

PRACTICE PROBLEMS

1. (L1 & L2) Complete and balance each chemical reaction:

a. $B_2O_3(s)$ + $Mg(s)$ $\rightarrow$

b. $Al_2O_3(s)$ + $HF(aq)$ + $NaOH(aq)$ $\rightarrow$

c. $PbO(s)$ + $C(s)$ $\rightarrow$

d. $CH_4(g)$ + $Cl_2(g)$ $\rightarrow$

e. $SiO_2(s)$ + $C(s)$ $\xrightarrow{2000^{\circ}C}$

f. $CaC_2(s)$ + $N_2(g)$ $\xrightarrow{heat}$

g. $CH_4(g)$ + $NH_3(g)$ $\xrightarrow{catalyst}$

2. **(L2)** What do rubies, sapphires, and corundum have in common?

3. **(L2)** Name some of the substances listed in this chapter that are commonly used as drying agents.

4. **(L1 & L2)** Write balanced chemical reactions for one method of preparation for each substance listed.

a. very pure elemental boron

b. very pure silicon

c. purification of bauxite ore

d. boron trifluoride

5. **(L2)** Name the allotropes of carbon and indicate which one is more stable.

6. **(L1)** One example of the class of compounds known as boranes consists of 85.7% boron and 14.3% hydrogen. What is the empirical formula of this compound?

7. **(L1)** If 0.132 grams of the boron compound discussed in the previous question is dissolved in 10.0 grams of benzene, the freezing point of the benzene is decreased by 1.07°C. What is the molar mass of this boron hydride? (K_f = 5.12°C/m for benzene.)

PRACTICE PROBLEM SOLUTIONS

1. a. $B_2O_3(s) + 3 Mg(s) \rightarrow 2 B(s) + 3 MgO(s)$

b. $Al_2O_3(s) + HF(aq) + NaOH(aq)$
$\rightarrow Na_3AlF_6(s) + H_2O(\ell)$

c. $PbO(s) + C(s) \rightarrow Pb(\ell) + CO(g)$

d. $CH_4(g) + 4 Cl_2(g) \rightarrow CCl_4(\ell) + 4 HCl(g)$

e. $SiO_2(s) + 3 C(s) \xrightarrow{2000°C} SiC(s) + 2 CO(g)$

f. $CaC_2(s) + N_2(g) \xrightarrow{heat} CaNCN(s) + C(s)$

g. $CH_4(g) + NH_3(g) \xrightarrow{catalyst} HCN(g) + 3 H_2(g)$

2. All three of these substances are different forms of aluminum oxide. Corundum is pure aluminum oxide; rubies contain chromium(III) as an impurity, and sapphires may be pure aluminum oxide or they may contain iron or titanium as impurities.

3. When soluble sodium silicate is treated with acid, the resulting precipitate, silica gel, can absorb up to 40% of its own weight of water. Zeolites are another silicon-containing species that absorb large amounts of water because of their structure.

4.

a. $2 \text{ BBr}_3(g) + 3 \text{ H}_2(g) \overset{\text{Ta}}{\rightarrow} 2 \text{ B}(s) + 6 \text{ HBr}(g)$

b. $\text{SiO}_2(s) + 2 \text{ C}(s) \overset{\text{heat}}{\rightarrow} \text{Si}(\ell) + 2 \text{ CO}(g)$

$\text{Si}(s) + 2 \text{ Cl}_2(g) \rightarrow \text{SiCl}_4(\ell)$

$\text{SiCl}_4(g) + 2 \text{ Mg}(s) \rightarrow 2 \text{ MgCl}_2(s) + \text{Si}(s)$

c. In the Bayer process, first SiO_2 and Al_2O_3 are dissolved in hot concentrated NaOH, impurities are removed by filtration, and the aluminum oxide is precipitated by the reaction:

$\text{H}_2\text{CO}_3(aq) + 2 \text{ NaAl(OH)}_4(aq)$
$\rightarrow \text{Na}_2\text{CO}_3(aq) + \text{Al}_2\text{O}_3(s) + 5 \text{ H}_2\text{O}(\ell)$

d. $\text{B}_2\text{O}_3(s) + 6 \text{ HF}(g) \rightarrow 2 \text{ BF}_3(g) + 3 \text{ H}_2\text{O}(\ell)$

5. The two allotropic forms of carbon are diamond and graphite. Graphite is slightly more stable than diamond.

6. Assume that 100.0 grams of compound is present and the calculate the moles of boron and hydrogen.

Moles of B = $\dfrac{85.7 \text{ g}}{10.81 \text{ g/mol}}$ = 7.93 moles

Moles of H = $\dfrac{14.3 \text{ g}}{1.008}$ = 14.2 moles

Find the mole ratio

Moles of B : moles of H :: $\dfrac{7.93}{7.83}$: $\dfrac{14.2}{7.83}$

:: 1 : 1.8

[This ratio should not be rounded back to 1:2!]

Moles of B : moles of H :: $\dfrac{10}{10}$: $\dfrac{18}{10}$

Moles of B : moles of H :: 10 : 18 : 5 : 9

The empirical formula is B_5H_9.

7. Use the freezing point depression equation to determine the molality of the benzene solution.

$$\Delta T = K_f m$$

molality of solute = $\dfrac{1.07^{\circ}C}{5.12^{\circ}C/m}$ = 0.20898 m

Using the defining equation for molality

Mole of solute = kilograms of solvent x molality

= 0.010 kg x 0.20898 m

= 0.0020898 moles

Now determine the molar mass.

Molar mass = $\dfrac{0.132 \text{ g}}{0.0020898 \text{ moles}}$

Molar mass = 63 g/mol

378

PRACTICE TEST (30 min.)

1. Complete and balance each of the following reactions:

a. $Al(s)$ + $O_2(g)$ $\rightarrow$

b. $CaF_2(s)$ + $H_2SO_4(\ell)$ $\rightarrow$

c. $Al(s)$ + $Cl_2(g)$ $\rightarrow$

d. $SnO_2(s)$ + $C(s)$ $\overset{heat}{\rightarrow}$

e. $PbS(s)$ + $O_2(g)$ $\rightarrow$

f. $Ni(s)$ + $CO(g)$ $\overset{50°C}{\rightarrow}$

g. $Ga(OH)_3(s)$ + $KOH(aq)$ $\rightarrow$

2. Write balanced chemical reactions for one method of preparation for each substance listed.

a. diborane, B_2H_6

b. boric carbide

c. Freon-11, $CFCl_3$

d. silane, SiH_4

3. Arrange the following oxides in order of increasing acidity when dissolved in water: B_2O_3, NO_2, Al_2O_3, and MgO.

4. What is the special significance of the compound diborane, B_2H_6?

5. List at least five different examples of silicate minerals.

6. Carbon and silicon are both in the same group of the periodic table and both form oxides with similar empirical formulas, CO_2 and SiO_2. Despite this, these two compounds are quite different from each other. Describe some of the major differences

379

between CO_2 and SiO_2, and explain why these differences exist.

7. Diborane, B_2H_6, is usually produced by the reaction of sodium borohydride with iodine. What volume of diborane gas measured at STP can be produced by the complete reaction of 1.00 kilograms of sodium borohydride?

CONCEPT TEST ANSWERS

1. allotropes
2. boron
3. aluminum oxide, Al_2O_3
4. gallium
5. borax
6. Because it's soft and weak unless alloyed with other elements.
7. more
8. battery plates or leaded glass
9. zeolites
10. silicon
11. tin and copper
12. carborundum
13. catenation
14. B_2H_6, diborane
15. refrigerant, dry ice

PRACTICE TEST ANSWERS

1. **(L1 & L2)**

a. $4 Al(s) + 3 O_2(g) \rightarrow 2 Al_2O_3(s)$

b. $CaF_2(s) + H_2SO_4(\ell) \rightarrow CaSO_4(s) + 2 HF(g)$

c. $2 Al(s) + 3 Cl_2(g) \rightarrow 2 AlCl_3(s)$

d. $SnO_2(s) + C(s) \xrightarrow{heat} Sn(s) + CO_2(g)$

e. $2 PbS(s) + 3 O_2(g) \rightarrow 2 PbO(s) + 2 SO_2(g)$

f. $Ni(s) + 4 CO(g) \xrightarrow{50^{\circ}C} Ni(CO)_4(\ell)$

g. $Ga(OH)_3(s) + KOH(aq) \rightarrow K^+ + [Ga(OH)_4]^{-1}$

2. **(L1 & L2)**

a. $2 NaBH_4(s) + I_2(s) \rightarrow B_2H_6(g) + 2 NaI(s) + H_2(g)$

b. $2 B_2O_3(s) + 4 C(s) \rightarrow B_4C(s) + 3 CO_2(g)$

380

c.
$$CCl_4(\ell) \quad + \quad HF(\ell) \quad \xrightarrow{\text{catalyst}} \quad CFCl_3(g) + HCl(g)$$

d. $SiCl_4(\ell) \quad + \quad 4\ NaH(s) \quad \rightarrow \quad SiH_4(g) \quad + 4\ NaCl(s)$

3. **(L1 & L2)** $MgO < Al_2O_3 < B_2O_3 < NO_2$

4. **(L1)** Diborane is the simplest compound of boron and hydrogen. It is also a classic example of the three-center bond. Each molecule of diborane contains two boron-to-hydrogen-to-boron bridges, which are best explained if the hydrogen 1s orbital overlaps with the hybrid orbitals from the borons to form a pair of molecular orbitals. This means that only two electrons are required to explain each bridge bond, even though three atoms are bonded together.

5. **(L2)** Portland cement and olivine are examples of orthosilicates; pyroxenes are typical metasilicates; the asbestos minerals are members of the group called amphiboles, and kaolinite is the example mentioned in the text to represent the large number of different clay minerals that are known. Feldspars and zeolites are examples of aluminosilicates.

6. **(L2)** Carbon dioxide is a simple, gaseous molecule with C=0 double bonds; silicon dioxide occurs is several different forms, but each form consists of infinite arrays of SiO_4 tetrahedra sharing corners. Because of this network structure, silicates are solids under normal conditions.

7. **(L1)** 296 liters

CHAPTER 24
THE CHEMISTRY OF THE NONMETALS: PERIODIC GROUPS 5A THROUGH 7A AND THE RARE GASES

LEARNING GOALS:

1. Learn about the chemical and physical properties, methods of preparation, compounds, and uses of the Group 5A elements. (Sec. 24.1)

2. Be familiar with the properties, methods of preparation, common compounds, and uses of the Group 6A elements. (Sec. 24.2)

3. Be familiar with the properties, methods of preparation, common compounds, and uses of the Group 7A elements. (Sec. 24.3)

4. Learn about the chemical and physical properties, methods of preparation, compounds, and uses of the rare gas elements. (Sec. 24.4)

IMPORTANT NEW TERMS:

aqua regia (24.1)
contact process (24.2)
hydrazine (24.1)
hypergolic (24.1)
orthophosphates (24.1)
Ostwald process (24.1)
ozone (24.2)
Raschig process (24.1)
superphosphate (24.1)

CONCEPT TEST

1. _____ is the most toxic as well as the most reactive allotrope of phosphorus.

2. The principal commercial use of arsenic and antimony is in the production of _____.

3. The Raschig process is used to prepare _____, which is one of the simplest catenated nitrogen compounds.

4. _____ is the blue, diamagnetic gas that is the primary allotrope of oxygen.

5. The major industrial source of oxygen gas is _____.

6. The repeating structural unit in elemental phosphorus consists of four phosphorus atoms that are arranged in space to form a(n) _____. This structure also serves as the basis for many compounds of phosphorus.

7. Match each of the nitrogen oxides on the left below with the best descriptive phase on the right. Use each phrase on the right only once.

_____ NO   a. laughing gas

_____ N_2O_4 b. brown air pollutant

_____ N_2O c. oxidizing agent in some rocket fuels

_____ NO_2 d. intermediate in the production of nitric acid by oxidation of ammonia

8. Commercially, _____ is the most important oxygen containing phosphorus acid. It is used for a variety of purposes, ranging from imparting corrosion resistance to giving a tart taste to soft drinks.

9. _____ is the acid used in such large quantities by the chemical industry that the amount of its production is used as an indicator of a nation's industrial strength.

10. The element _____ is used in photocopying machines in order to form the image that is transferred to the paper.

11. _____ is the compound of oxygen and hydrogen sometimes used as a bleaching or sterilizing agent.

12. Hypo, which has the formula _____, is used as a fixer when developing photographic films, that is, the hypo dissolves the unreacted silver salts and prevents further light from changing the image on the film.

13. Most of the chlorine produced is used commercially for _____.

14. Most of the hydrogen fluoride produced is used to make _____, a relatively rare mineral that is necessary for aluminum production.

15. The normal boiling point of liquid _____ is only 4.2 K, and so it is the coldest liquid refrigerant available.

PRACTICE PROBLEMS

1. (L3) Match each of the halogens on the left below with the best descriptive phase on the right.

_____ bromine   a. purple-black solid

_____ fluorine   b. yellow-green gas

_____ iodine   c. colorless gas

_____ chlorine   d. red-brown liquid

2. (L3) Which halogen is the strongest oxidizing agent?

a. F_2 b. I_2 c. Br_2 d. Cl_2 Answer: _____

3. **(L3)** Most of the fluorine produced is used commercially for _____.

4. **(L1)** Which of the following oxides is least likely to be soluble in basic solution?

a. As_4O_6 b. Bi_2O_3 c. Sb_4O_6 Answer: _____

5. **(L1 thru L3)** Complete and balance each of the following equations:

a. $Fe(s)$ + $Sb_2S_3(s)$ $\rightarrow$

b. $N_2H_4(aq)$ + $O_2(g)$ $\rightarrow$

c. $P_4O_{10}(s)$ + $H_2O(\ell)$ $\rightarrow$

d. $As_4O_6(s)$ + $HCl(aq)$ $\rightarrow$

e. $H_2SeO_3(aq)$ + $SO_2(g)$ + $H_2O(\ell)$ $\rightarrow$

f. H_2S + SO_3 (-78 C, no water) $\rightarrow$

g. $NaCl(s)$ + $H_2SO_4(aq)$ + $MnO_2(s)$ $\rightarrow$

h. $Br^-(aq)$ + $Cl_2(aq)$ $\rightarrow$

1. $Cl_2(g)$ + $OH^-(hot\ aq)$ $\rightarrow$

6. **(L1 thru L3)** Write balanced chemical reactions for one method of preparation for each substance listed.

a. phosphine gas

b. sulfuric acid

c. hydrogen chloride

d. phosphoric acid

385

7. (**L1 thru L4**) Use VSEPR Theory to predict the structure of the following compounds:
a. NF_3 b. PF_5 c. H_2Se d. XeF_4 e. ClF_3

8. (**L2**) You are already familiar with the hydrides of the first two elements of Group VI, H_2O and H_2S. The hydride of the third member of this group, H_2Se, is usually not as well known. Based on your general chemical knowledge and periodic correlations, predict as much as you can about the chemistry and structure of H_2Se.

9. (**L3**) The average person in this country uses 380 liters of water a day. How many grams of sodium fluoride would be necessary to fluoridate this much water to a level of 1 part per million by mass fluoride ion?

PRACTICE PROBLEM SOLUTIONS

1. d.- bromine   c.- fluorine
   a.- iodine   b.- chlorine

2. F_2, fluorine

3. processing uranium for nuclear fuel plants

4. b. Bi_2O_3 is the most basic of these salts

5. a. $3 Fe(s) + Sb_2S_3(s) \rightarrow 3 FeS(s) + 2 Sb(s)$

   b. $N_2H_4(aq) + O_2(g) \rightarrow N_2(g) + 2 H_2O(\ell)$

   c. $P_4O_{10}(s) + 6 H_2O(\ell) \rightarrow 4 H_3PO_4(aq)$

   d. $As_4O_6(s) + 12 HCl(aq) \rightarrow 4 AsCl_3(aq) + 6 H_2O(\ell)$

   e. $H_2SeO_3(aq) + 2 SO_2(g) + H_2O(\ell)$
   $\rightarrow Se(s) + 2 H_2SO_4(aq)$

   f. $H_2S + SO_3$ (-78 C, no water)$\rightarrow H_2S_2O_3$

   g. $2 NaCl(s) + 2 H_2SO_4(aq) + MnO_2(s)$
   $\rightarrow Na_2SO_4(aq) + MnSO_4(aq) + 2 H_2O(\ell) + Cl_2(g)$

   h. $2 Br^-(aq) + Cl_2(aq) \rightarrow 2 Cl^-(aq) + Br_2(aq)$

   1. $3 Cl_2(g) + 6 OH^-$ (hot aq)
   $\rightarrow ClO_3^-(aq) + 5 Cl^-(aq) + 3 H_2O(\ell)$

6. a. $P_4(s) + 3 KOH(aq) + 3 H_2O(\ell)$
   $\rightarrow PH_3(g) + 3 KH_2PO_2(aq)$

   b. The Contact Process

   $S(s) + O_2(g) \xrightarrow{heat} SO_2(g)$

   $SO_2(g) + 1/2 O_2(g) \xrightarrow{catalyst} SO_3(g)$

   $SO_3(g) + H_2SO_4(\ell) \rightarrow H_2SO_4(\ell)$

   c. $2 NaCl(s) + H_2SO_4(\ell) \rightarrow Na_2SO_4(s) + 2 HCl(g)$

d. $Ca_3(PO_4)_2(s) + 3\ H_2SO_4(aq)$
$$\longrightarrow 2\ H_3PO_4(aq) + 3\ CaSO_4(s)$$
or for very pure acid

$$P_4(s) \quad + \quad 5\ O_2(g) \overset{heat}{\longrightarrow} P_4O_{10}(s)$$

$$P_4O_{10}(s) \quad + \quad 6\ H_2O(\ell) \longrightarrow \quad 4\ H_3PO_4(aq)$$

7.

electron pair geometry	number of lone pairs	molecular geometry
a. tetrahedron	one	trigonal pyramid
b. trigonal bipyramid	none	trigonal bipyramid
c. tetrahedron	two	bent (109°)
d. octahedron	two	square planar
e. trigonal bipyramid	two	T-shaped

8. VSEPR Theory would suggest that H_2Se should have a bent structure with a bond angle somewhat smaller than the normal tetrahedral angle (109.5°). The electronegativity of selenium is much less than that of oxygen, and this suggests that hydrogen bonding will not be important in hydrogen selenide. This would lead one to suggest that hydrogen selenide is a gas under normal conditions of temperature and pressure. By analogy with hydrogen sulfide, you should expect hydrogen selenide to be a weak acid, and in fact it should probably be somewhat stronger than hydrogen sulfide.

9. First, determine the mass of water used.

$$\text{grams of water} = \frac{380\ L\ water}{} \times \frac{1000\ mL}{1\ L} \times \frac{1\ gram}{1\ mL}$$

grams of water = 3.80×10^5 grams

Next, find the moles of fluoride ion needed.

$$\text{mol of } F^- = \frac{3.80 \times 10^5\ g\ water}{} \times \frac{1\ gram\ F^-}{1 \times 10^6\ g\ water} \times \frac{1\ mol\ F^-}{19.0\ g}$$

388

mol of $F^- = 0.020$ mole

Each mole of fluoride ion requires one mole of NaF.

grams of NaF = 0.020 mol F^- x $\dfrac{1 \text{ mol NaF}}{1 \text{ mol } F^-}$ x $\dfrac{42.0 \text{ g}}{1 \text{ mol NaF}}$

= <u>0.8 grams of NaF needed</u>

PRACTICE TEST (45 min.)

1. **(L1)** Which of the following elements is <u>least</u> likely to form compounds in which it has the +5 oxidation state?

a. bismuth   b. arsenic   c. antimony

2. **(L1)** Arrange the following oxides in order of increasing acidity when in aqueous solution: N_2O_5, Bi_2O_3, and Sb_4O_6.

3. **(L1)** Match each of the phosphate salts on the left below with the best descriptive phase on the right. Use each phrase on the right only once.

_____ $Na_5P_3O_{10}$ a. superphosphate fertilizer

_____ $CaHPO_4$ b. phosphate detergent builder

_____ Na_2HPO_4 c. Kraft process for pasteurized cheese

_____ $Ca(H_2PO_4)_2$ d. toothpaste abrasive

4. **(L3)** Name each of the following acids and salts:

A. $NaClO_2$ _____   B. HOF _____

C. $KBrO_3$ _____   D. HOI_3 _____

5. **(L1 and L4)** Use VSEPR Theory to predict the structure of the following compounds:

a. BeF_2 b. $AsCl_3$ c. OF_2 d. XeF_2 e. I_3^-

389

6. (L1 thru L4) Complete and balance each of the
following equations:

a. $As_4O_6(s)$ + $NaOH(aq)$ $\rightarrow$

b. $Xe(g)$ + F_2 + sunlight $\rightarrow$

c. $Mg(s)$ + $N_2(g)$ $\rightarrow$

d. $NaNO_2(aq)$ + $HCl(aq)$ $\rightarrow$

e. $Mg(s)$ + $HCl(g)$ $\rightarrow$

f. $NH_4NO_3(s)$ + heat$(>300°C)$ $\rightarrow$

g. $CaF_2(s)$ + $H_2SO_4(\ell)$ $\rightarrow$

h. $CuS(s)$ + $O_2(g)$ $\rightarrow$

i. $Cl_2(g)$ + OH^-(cold aq) $\rightarrow$

j. $BaO_2(s)$ + $H^+(aq)$ $\rightarrow$

7. (L1 and L2) Write balanced chemical reactions
for one method of preparation for each substance
listed.

a. arsine gas
b. nitric acid
c. oxygen gas
d. hydrazine

8. (L1) In 1966 a new compound was prepared
consisting only of phosphorus and fluorine and
having the empirical formula PF_2. It was observed
that a sample of this compound having a mass of
0.217 grams would exert a pressure of 244 mmHg at
25°C in a container having a volume of 120 mL. What
is the molecular formula of this compound?

CONCEPT TEST ANSWERS

1. white phosphorus   2. automobile batteries
3. hydrazine, N_2H_4 4. ozone
5. fraction distillation of the atmospheric gases

6. tetrahedron
7. d.- NO, c.- N_2O_4, a.- N_2O, and b.- NO_2
8. orthophosphoric acid
9. sulfuric acid   10. selenium
11. hydrogen peroxide   12. $Na_2S_2O_3 \cdot 5H_2O$
13. production of organic chemicals
14. cryolite   15. helium

PRACTICE TEST ANSWERS

1. Bismuth (The stability of the higher oxidation states decreases as you go down this group.)

2. $Bi_2O_3 < Sb_4O_6 < N_2O_5$ (These elements become more metallic as you go down the group.)

3. b.- $Na_5P_3O_{10}$ d.- $CaHPO_4$

   c.- Na_2HPO_4 a.- $Ca(H_2PO_4)_2$

4. a. $NaClO_2$ - sodium chlorite
   b. HOF - hypofluorous acid
   c. $KBrO_3$ - potassium bromate
   d. HOI_3 - iodic acid

5.

electron pair geometry	number of lone pairs	molecular geometry
a. linear	none	linear
b. tetrahedron	one	trigonal pyramid
c. tetrahedron	two	bent (109°)
d. trigonal bipyramid	three	linear
e. trigonal bipyramid	three	linear

6.a. $As_4O_6(s) + 12\ NaOH(aq) \rightarrow 4\ Na_3AsO_3(aq) + 6\ H_2O(\ell)$

b. $Xe(g) + F_2 + sunlight \rightarrow XeF_2(s)$

c. $3\ Mg(s) + N_2(g) \rightarrow Mg_3N_2(s)$

d. $NaNO_2(aq) + HCl(aq) \rightarrow NaCl(aq) + HNO_2(aq)$

e. $Mg(s) + 2 HCl(g) \rightarrow MgCl_2(s) + H_2(g)$

f. $NH_4NO_3(s) + heat(>300°C)$
$$\rightarrow 2 N_2(g) + O_2(g) + 4 H_2O(g)$$

g. $CaF_2(s) + H_2SO_4(\ell) \rightarrow CaSO_4(s) + 2 HF(g)$

h. $2 CuS(s) + 3 O_2(g) \rightarrow 2 CuO(s) + 2 SO_2(g)$

i. $Cl_2(g) + 2 OH^-(cold\ aq)$
$$\rightarrow OCl^-(aq) + Cl^-(aq) + H_2O(\ell)$$

j. $BaO_2(s) + 2 H^+(aq) \rightarrow H_2O_2(aq) + Ba^{2+}(aq)$

7. a. $H_3AsO_4(aq) + 4 H_2(g) \rightarrow 4 H_2O(\ell) + AsH_3(g)$

b. $2 NaNO_3(s) + H_2SO_4(aq) \rightarrow 2 HNO_3(aq) + Na_2SO_4(s)$

An alternate method of synthesis, called the Ostwald Process, could also be used to answer this question.

b.
$NH_3(g) + 5/4 O_2(g) \xrightarrow{Pt} NO(g) + 3/2 H_2O(g)$

$NO(g) + 1/2 O_2(g) \rightarrow NO_2(g)$

$NO_2(g) + 1/3 H_2O(\ell) \rightarrow 2/3 HNO_3(aq) + 1/3 NO(g)$

c. Oxygen is normally prepared by fractional distillation of air, but it can also be chemically prepared by the reaction

$2 KClO_3(s) \xrightarrow{catalyst\ \&\ heat} 2 KCl(s) + 3 O_2(g)$

d. (The Raschig Process)
$2 NH_3(aq) + NaOCl(aq) \rightarrow N_2H_4(aq) + NaCl(aq) + H_2O(\ell)$

8. (L1) molar mass = 138 g/mole; formula = P_2F_4

CHAPTER 25
THE TRANSITION ELEMENTS
AND THEIR CHEMISTRY

LEARNING GOALS:

1. Understand the special chemical and physical properties of the transition metals, especially those that are related to the electronic configurations of these elements. (Sec. 25.1)

2. Be familiar with the commercial methods used for the production and purification of the transition metals. (Sec. 25.2)

3. Know how to name coordination compounds and also to identify the possible types of isomerism that may occur in this class of compounds. (Secs. 25.3 & 25.4)

4. Be able to use a modern theory of chemical bonding to discuss the bonding in coordination compounds and to explain the magnetic properties and colors of coordination compounds. (Secs. 25.4 & 25.6)

IMPORTANT NEW TERMS:

ferromagnetic (25.1)
basic oxygen furnace (25.2)
bidentate (25.3)
chelate (25.3)
chemical leaching (25.2)
chiral center (25.4)
crystal field theory (25.5)
enantiomers (25.4)
flotation (25.2)
functional isomers (25.4)
gangue (25.2)
geometric isomerism (25.4)
high spin complexes (25.5)
hydrometallurgy (25.2)
lanthanide contraction (25.1)

CONCEPT TEST

1. Transition metal ions are most often found in high oxidation states when they are combined with nonmetals such as _____ or _____.

2. _____ and _____ are the most commonly observed oxidation numbers for the d-block elements, although a wide variety of other values are also observed.

3. Most of the d-block and f-block elements have unpaired electrons in the ground state and are said to be _____.

4. Several observations, including the fact that radii of the sixth period transition elements are almost identical to those of the corresponding fifth period transition elements, are explained by the _____.

5. _____ is the use of high temperatures to recover metals from their ores, and _____ is the use of low temperature,

aqueous solutions for the same purpose.

6. The combination of calcium silicate and metal oxides that floats on top of the liquid iron in a blast furnace is called _____.

7. Both paramagnetic and ferromagnetic materials contain unpaired electrons, but in a ferromagnet large arrays of unpaired electrons interact to form a group of aligned magnets called a(n) _____.

8. The Lewis bases bonded to the central metal ion in a coordination compound are called _____.

9. When a ligand has more than one Lewis base site, it can bond to a metal ion in each of these sites to form a type of complex called a(n) _____.

10. Give the oxidation number of the metal ion in each complex.

a. $[PtCl_4]^{2-}$ _____ b. $[Co(NH_3)_4Cl_2]^+$ _____

11. In naming a coordinating compound, if the complex ion is an anion, the suffix _____ replaces the normal ending of the metal name.

12. Beside each prefix, write the corresponding numerical value.

a. tris _____ b. tetrakis _____ c. bis _____

13. In naming a coordination compound, if the

ligand is an anion with a name that normally ends in -ide, the ending is changed to _____.

14. _____ are molecules with the same stoichiometry but different atomic arrangements. When two isomers are nonsuperimposable mirror images of each other, the compounds are called

_____.

15. The color and magnetic properties of the transition metal ion in coordination compounds is explained in terms of the energy shifts of the ___ orbitals. These energy effects cause this set of orbitals to split into two or more sets that have different energies.

16. The _____ series results when ligands are arranged in order of their ability to cause orbital splittings.

STUDY HINTS:

1. It was noted earlier that students sometimes confuse ammonia and ammonium. Also be careful to distinguish between the names ammine, which represents ammonia acting as a ligand, and amine, which is an organic functional group.

2. Remember that in some coordination compounds, some of the species are covalently bonded to the central metal ion, and others are bonded ionically. When such a compound is dissolved in water, the ligands that are coordinated usually remain bonded to the central metal ion, and so are less likely to

react. The brackets indicate which species are coordinated and which are not. Be sure to take this into consideration when trying to predict the reactions and structures of coordination compounds.

3. Molecular models can be very helpful in the study of isomerism.

4. When naming coordination compounds, if the coordinated species is an anion the metal name must end in -ate. For a few metals the name is also changed to the Latin form. Some examples of this change are iron becoming ferrate, silver becoming argentate, and gold becoming aurate. Watch for examples of these changes in the text and in this chapter.

5. If the name of a coordination compound contains a double vowel it is difficult to pronounce, and there is a temptation to drop one of the vowels. For example, hexaamminecobalt(III) ion could become hexamminecobalt(III). Try to avoid this unless your professor specifically indicates it is acceptable.

PRACTICE PROBLEMS

1. (L1) Give spectroscopic notation for the electronic configurations of the following transition metal ions and in each case indicate whether or not it is paramagnetic.

a. Zn^{2+} _____

b. Re^{2+} _____

c. Mn^{2+} _____

d. Zr^{4+} _____

2. (L3) Name the following ligands:

a. Br^- _____   b. CN^- _____

c. en _____   d. NCS^- _____

e. F^- _____   f. CO _____

3. **(L4)** Which of the following ions is <u>least</u> likely to produce highly colored compounds and why?

a. Co^{2+} b. Sc^{3+} c. Cd^{2+} a. Mo^{3+}

4. **(L3)** Name the following coordination compounds

a. $[Ti(H_2O)_6]^{3+}$ _____

b. $[AgF_4]^-$ _____

c. $[Co(CO)_4]^-$ _____

d. $[Pt(NH_3)_2Br_2]$ _____

e. $[Cu(NH_3)_4]SO_4$ _____

f. $[W(CN)_8]^{3-}$ _____

5. **(L3)** Write the formulas for each of the following compounds or ions:

a. tris(ethylenediamine)cobalt(III) _____

b. tris(oxalato)ferrate(III) _____

c. hexaamminecobalt(III) ion _____

d. hexafluoroantimonate(V) ion _____

6. **(L3)** Draw the geometric isomers of the following compounds and label each as a cis or trans isomer.

a. $[Pt(NH_3)_2Br_2]$ b. $[Co(NH_3)_4Br_2]^+$

7. **(L2)** Explain the fundamental chemistry of a basic oxygen furnace, using a few typical reactions to demonstrate the discussion.

PRACTICE PROBLEM SOLUTIONS

1. a. $[Ar]3d^{10}$ diamagnetic
   b. $[Xe]4f^{14}5d^5$ paramagnetic
   c. $[Ar]3d^5$ paramagnetic
   d. $[Kr]4d^0$ diamagnetic

2. a. Br^- bromo   b. CN^- cyano

   c. en   ethylenediamine   d. NCS^- thiocyanato

   e. F^- fluoro   f. CO   carbonyl

3. For coordination compounds, the color usually results from the splitting of the set of d orbitals. Electrons in these orbitals shift from one set to another, and the energy changes correspond to the colors we observe. Scandium(III) has no d electrons and so would not be expected to produce color in this way. Cadmium(II) has ten d electrons, just enough to fill all of the d orbitals in the valence shell. Since there are not empty spots left for the shift of the d electrons, this ion would not be expected to produce colorful compounds. The other two ions listed have d electrons and the d orbitals are not filled, and so they are expected to be colored.

4. a. hexaaquotitanium(III)
   b. tetrafluoroargentate(III)
   (See helpful hint #4.)
   c. tetracarbonylcobalt(-1)

399

d. diamminedibromoplatinum(II)
e. tetraamminecopper(II) sulfate
f. octacyanotungsten(V)

5. a. $[Co(en)_3]^{3+}$

   b. $[Fe(C_2O_4)_3]^{3-}$

   c. $[Co(NH_3)_6]^{3+}$

   d. $[SbF_6]-$

6. a.

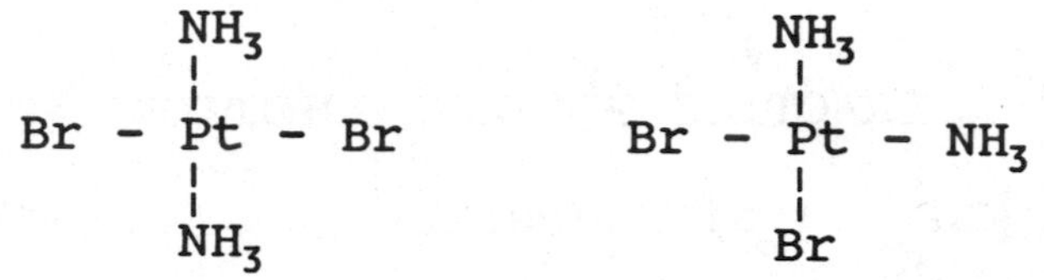

            trans isomer   cis isomer

b.

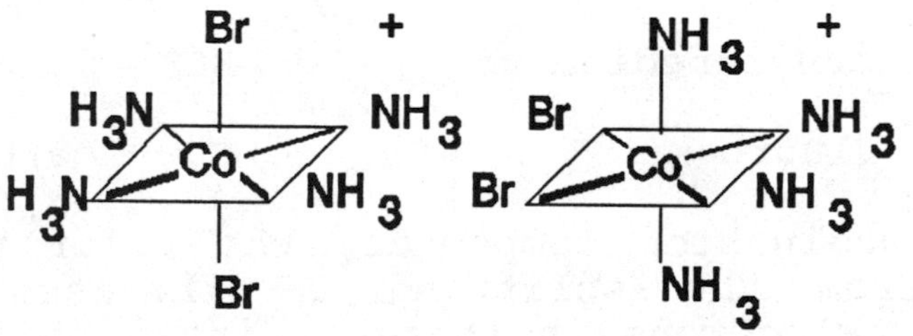

            trans isomer   cis isomer

7. The major purpose of a basic oxygen furnace is to remove nonmetallic impurities from pig iron. When oxygen gas is blown through the molten iron, phosphorus, sulfur, and carbon are oxidized, as shown in the following reactions:

$$4\ P(s) + 5\ O_2(g) \rightarrow 2\ P_2O_5(s)$$

$$C(s) + O_2(g) \rightarrow CO_2(g)$$

$$S(s) + O_2(g) \rightarrow SO_2(g)$$

These acidic oxides react with basic oxides, such

400

as CaO, that are added or used to line the furnace.

$$P_2O_5(s) \quad + \quad 3\ CaO(s) \quad \rightarrow \quad Ca_3(PO_4)_2(s)$$

The silicate minerals present, called gangue, are removed by reacting with the lime that results from heating the limestone.

$$SiO_2(s) \quad + \quad CaO(s) \quad \rightarrow \quad CaSiO_3(s)$$

PRACTICE TEST (30 Min.)

1. Write the electronic configuration for each of the following transition metal ions and indicate in each case whether or not the ion is paramagnetic.

a. Mo^{3+} b. Co^{2+} c. Cd^{2+} d. Cu^+

2. Discuss the sequence of reactions that occur in a blast furnace and are responsible for the conversion of iron ore into iron.

3. For each of the following compounds draw all of the geometric isomers and stereoisomers that are possible.

a. $[Co(en)Br_4]^-$ b. $[Cr(NH_3)_4Br_2]^+$ c. $[Co(en)_3]^{3+}$

4. Name the following coordination compounds

a. $[Ag(NH_3)_2]^+$ b. $[Co(en)_3]_2(SO_4)_3$

c. $K_2[MnF_6]$ d. $[Co(NH_3)_5Cl]Cl_2$

e. $[SbCl_6]^{3-}$ f. $K_2[NiF_6]$

5. Write the formulas for each of the following compounds or ions:

a. tetrabromoplatinum(II) ion
b. hexaaquoiron(III) ion
c. potassium diiodoargentate(I)
d. tetraaquodichlorochromium(III) ion

6. A certain coordination compound has the empirical formula $Cr(NH_3)_4Cl_3$, but when silver

nitrate is added to a solution of this species, only one mole of silver chloride is formed per mole of chromium in the solution. If possible, suggest an explanation for this observation.

7. Gold and platinum are among the most dense elements known. Suggest an explanation in terms of atomic structure for this observation.

CONCEPT TEST ANSWERS

1. fluorine and oxygen
2. +2 and +3   3. paramagnetic
4. lanthanide contraction
5. pyrometallurgy, hydrometallurgy
6. slag   7. domain
8. ligands   9. chelate
10. platinum(II), cobalt(III) 11. -ate
12. a. tris 3   b. tetrakis 4   c. bis 2
13. -o
14. isomers, optical isomers
15. d orbitals   16. spectrochemical

PRACTICE TEST ANSWERS

1. (L1) a. $[Kr]4d^3$ paramagnetic

   b. $[Ar]3d^7$ paramagnetic

   c. $[Kr]4d^{10}$ diamagnetic

   d. $[Ar]3d^{10}$ diamagnetic

2. (L2) The coke burns and produces carbon monoxide.

$$2 C(s, coke) + O_2(g) \rightarrow CO(g)$$

The carbon monoxide reduces the iron oxide to the metal.

$$Fe_2O_3(s) + 3 C(s) \rightarrow 2 Fe(\ell) + 3 CO_2(g)$$

Much of the carbon dioxide produced is reduced by the coke to form more carbon monoxide for the above reaction.

$$CO_2(g) + C(s) \rightarrow 2 CO(g)$$

402

3. **(L3)** a. Neither cis-trans nor stereoisomers are possible for this compound, that is, it exists in only one form.

b. This compound does have a pair of cis-trans isomers, as shown below.

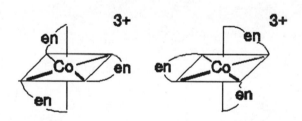

trans isomer   cis isomer

c. This compound has no cis-trans isomers, but it does have a pair of stereoisomers, as shown below.

4. **(L3)** a. diamminesilver(I) ion
   b. tris(ethylenediamine)cobalt(III) sulfate
   c. potassium hexafluoromanganate(IV)
   d. pentaamminechlorocobalt(III) chloride
   e. hexachloroantimonate(III) ion
   f. potassium hexafluoronickelate(IV)

5. (**L3**) a. $[PtBr_4]^{2-}$

b. $[Fe(H_2O)_6]^{3+}$

c. $K[AgI_2]$

d. $[Cr(H_2O)_4Cl_2]^+$

6. (**L3**) The chloride ions that are coordinated to the cobalt will not precipitate, but those outside the coordination sphere will. This suggests that the reaction is

$$[Co(NH_3)_4Cl_2]Cl(aq) + Ag^+(aq)$$
$$\rightarrow [Co(NH_3)_4Cl_2]^+ + AgCl(s)$$

7. (**L4**) The lanthanide contraction causes the radii of transition elements in the sixth period to be about the same size as the their counterparts in the fifth period. This relatively small size and the higher mass causes the densities to be unusually high.

CHAPTER 26
ORGANIC CHEMISTRY

LEARNING GOALS:

1. Know how to write structural formulas for and also name the simple alkanes. (Sec. 26.2)

2. Be familiar with the nomenclature, methods of synthesis, and some common reactions of alcohols and ethers. (Secs. 26.4 & 26.5)

3. Be able to write names for the common alkenes and alkynes, in addition to knowing some of the common reactions and preparation methods for these compounds. This will include knowing how to use the Markovnikov rule to predict the products of alkene addition reactions. (Sec. 26.6)

4. Be familiar with some of the nomenclature and reactions of aromatic compounds, especially benzene. (Sec. 26.7)

5. Be familiar with the nomenclature, methods of synthesis, and common reactions of alkyl halides, including the use of Grignard reagents in organic synthesis. (Sec. 26.8)

6. Understand the nomenclature and reactions of the carboxylic acids and esters. (Sec. 26.9)

7. Be able to write names for carbonyl compounds, such as aldehydes and ketones, in addition to knowing some of the common reactions and preparation methods for these compounds. (Sec. 26.10)

8. Be familiar with the nomenclature, methods of synthesis, and common reactions of amines and amides. (Sec. 26.11)

9. Many of the compounds discussed so far are the result of the refining of petroleum, so it's important to have a basic understanding of the major types of refinery processes. (Sec. 26.12)

IMPORTANT NEW TERMS:

addition reaction (26.4)
alcohol (26.4)
aldehyde (26.10)
alkane (26.1)
alkene (26.6)
alkyl halide (26.8)
alkyl group (26.2)
alkylation (26.12)
alkyne (26.6)
amide (26.11)
amine (26.11)
aromatic compound (26.7)
carbonyl group (26.3)
carboxylic acid (26.9)
catalytic cracking (26.12)
catalytic reforming (26.12)
common name (26.2)
elimination reaction (26.4)
ester (26.9)
ether (26.5)
functional group (26.3)
Grignard reagent (26.8)
hydrogenation (26.6)
ketone (26.10)
Markovnikov's rule (26.6)
saponification reaction (26.9)
saturated hydrocarbon (26.2)
soap (26.9)
steam cracking (26.12)
structural isomer (26.2)
substitution reaction (26.7)
systematic name (26.2)
unsaturated hydrocarbon (26.2)

CONCEPT TEST

1. Circle the unsaturated hydrocarbon.

A. $CH_3CH_2CH_3$ B. CH_4 C. CH_3CH_3 D. $CH_2=CH_2$

2. The hydrocarbon _____ is the principal constituent of natural gas.

3. _____ is the name of the alcohol found in alcoholic beverages.

4. Name the compounds having the structural formulas given.

A. $CH_3CH_2CH_3$ B. $CH_3CH_2CH_2CH_3$

C. $CH_3CH_2CH_2CH_2CH_2CH_3$ D. CH_3CH_3

5. The systematic names of alkanes are based on the number of carbon atoms in the _____ carbon chain.

6. Which of the following compounds can be used as a reducing agent for certain organic compounds?

A. $LiAlH_4$ B. $KMnO_4$

C. $Na_2Cr_2O_7$ D. Ag^+

7. _____ is the name of the compound commonly known as automobile antifreeze.

8. Ethylene and propylene are prepared industrially by heating the hydrocarbons found in natural gas and petroleum, a process called _____.

9. _____ is the acid responsible for the sour taste of vinegar, and _____ is responsible for the unpleasant odor of rancid butter.

10. The odor of many fruits, such as oranges, apples, bananas, and pineapples, is produced by individual compounds that are in the general class of compounds called _____.

11. The hydrolysis of esters is called a(n) _____ reaction, because in some cases the product is a soap.

12. _____ are the class of compounds responsible for the odor of decaying fish or decaying flesh.

13. Match the organic functional groups in the list

on the right below with the appropriate suffix or ending from organic nomenclature that corresponds to that functional group.

_____ -al   A. ketone

_____ -ene   B. aldehyde

_____ -yne   C. carbon-carbon double bond

_____ -oic acid   D. alcohol

_____ -ol   E. organic acid

_____ -one   F. carbon-carbon triple bond

14. When two substituents are attached to adjacent carbons on a benzene ring, they are said to be _____ ; when two substituents are attached to carbons that are separated by two carbon atoms on the benzene ring, they are said to be _____.

15. The refinery process called _____ is used to break down large alkanes into smaller, branched-chain alkanes; the process called _____ converts alkanes to aromatic compounds.

STUDY HINTS:

1. Organic chemistry requires somewhat different study techniques than most of the material previously encountered in this course. The temptation will probably be to attempt to memorize each separate reaction listed in this chapter. Even though the number of reactions given here is only a small fraction of the number that would be encountered in a full course in organic chemistry, there are still too many to memorize effectively in this way. It is essential to attempt to organize the information before memorizing it. The concept of a functional group provides a start in this direction. Remember that learning the reactions of a single functional group actually means learning the chemistry of a class of compounds that may be

quite large. Next, look for similarities and differences in various reaction types. As more different reactions are learned, it will become obvious that the relationships learned previously will make it easier to memorize the new material. This will not only make the new chemistry easier to learn but the review should also help to better remember the old material.

2. Many problems in organic chemistry may require visualization of molecules in three dimensions, a skill best learned by working with molecular models. Except for those who plan to take an organic course in the future, it probably isn't feasible to purchase a set of molecular models but try to take advantage of every opportunity to use the set in class or use models borrowed from a friend.

3. Synthetic problems can be especially difficult because they require an understanding of how to go from one type of compound to another. Simply memorizing reactions without seeing relationships is unlikely to produce this skill. One way to improve this ability is to draw a "reaction map." Take a large piece of paper and write the word "alkanes" in the center of the page. A few inches away, write the word "alkenes." Now write down all of the reactions that allow you to go from one to another of these classes of compounds. Next add other functional groups, in each case writing the reactions that connect the new type of compound to all of the functional groups already on the page. When attempting to design a synthetic route, visualize the piece of paper, and think about how the reactions lead from the starting material to the final product desired. This technique can be especially useful if you take further courses in organic chemistry.

PRACTICE PROBLEMS

1. (L6) Which of the following compounds can form an acid when it is oxidized?

a. 2-pentanol b. butanone c. hexanal d. 3-pentanol

2. **(L2)** Which of the following types of alcohol does not react with oxidizing agents?

a. primary alcohol
b. secondary alcohol
c. tertiary alcohol

3. **(L1, L2, L3, L4, L6, and L8)** Write systematic names for each of the following compounds.

a. CH_3-O-CH_3 _____

b. $CH_3CH=CHCH_3$ _____

c. CH_3CH-OH _____
       |
      CH_3

d.
        O
        ‖
$CH_3CH_2CH_2CCH_3$ _____

e.
    NH_2

_____

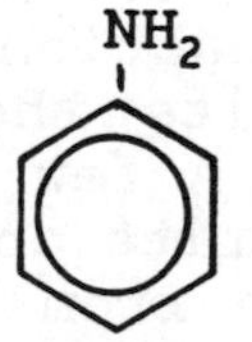

f.
    CH_3

        CH_3 _____

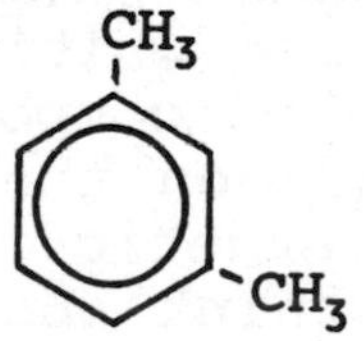

4. **(L1)** Draw all of the structural isomers having the formula C_6H_{14}.

5. (L1) Name each of the structural isomers you drew in the previous question.

6. (L1, L3, L4, L5, L6, and L8) Draw the structure of each of the following compounds:

a. 1-pentyne   b. 1,2-diaminoethane

c. oxalic acid   d. 2-hexanol

e. propanoic acid   f. ethyl acetate

g. 2-butanone   h. o-dichlorobenzene

7. (L1 thru L8) Write the correct product(s) to complete the following equations.

a. $CH_3CH_2CH_2OH$ + H_2SO_4 $\rightarrow$

$$
\begin{array}{c}
\quad\quad\quad O \\
\quad\quad\quad \parallel \quad\quad Na_2Cr_2O_7 \\
b.\ CH_3CH_2CH \quad \rightarrow
\end{array}
$$

$$
\begin{array}{c}
\quad\quad\quad OH \\
\quad\quad\quad | \quad\quad\quad KMnO_4 \\
c.\ CH_3CH_2CHCH_3 \quad \rightarrow
\end{array}
$$

d. $CH_3CH=CHCH_2CH_3 + H_2(g) \xrightarrow{\text{metal catalyst}}$

e. $CH_2=CHCH_2CH_3 + HCl \rightarrow$

f. $CH_3CH_2\overset{\overset{\displaystyle O}{\|}}{C}-OH \xrightarrow[\text{(b) } H_2O]{\text{(a) } LiAlH_4}$

PRACTICE PROBLEM SOLUTIONS

1. c. hexanal

2. c. tertiary alcohol

3. a. dimethylether
   b. 2-butene (note cis and trans isomers are possible)
   c. 2-propanol   d. 2-pentanone
   e. aniline   f. **m**-dimethylbenzene

4. $CH_3CH_2CH_2CH_2CH_2CH_3$

    n-hexane

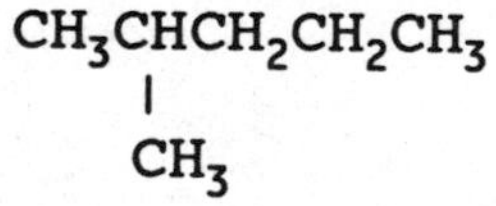

$CH_3\overset{\displaystyle }{\underset{\underset{\displaystyle CH_3}{|}}{C}}HCH_2CH_2CH_3$

2-methylpentane

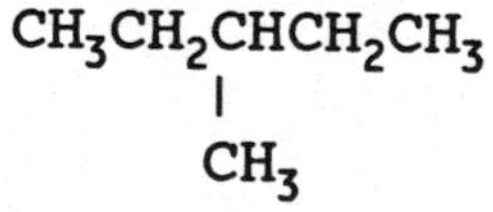

$CH_3CH_2\overset{\displaystyle }{\underset{\underset{\displaystyle CH_3}{|}}{C}}HCH_2CH_3$

3-methylpentane

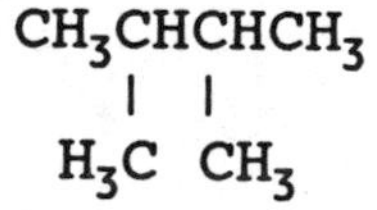

$CH_3\overset{\displaystyle }{\underset{\underset{\displaystyle H_3C}{|}}{C}}H\overset{\displaystyle }{\underset{\underset{\displaystyle CH_3}{|}}{C}}HCH_3$

2,3-dimethylbutane

$CH_3\overset{\overset{\displaystyle CH_3}{|}}{\underset{\underset{\displaystyle CH_3}{|}}{C}}CH_2CH_3$

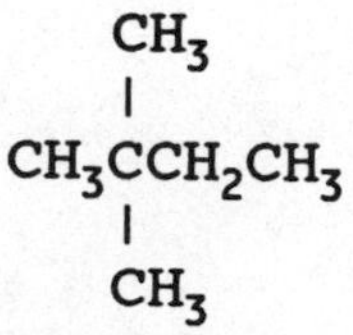

2,2-dimethylbutane

5. see above answer

412

6. a. $HC \equiv CCH_2CH_2CH_3$

   b. $NH_2CH_2CH_2NH_2$

c.
$$\underset{\displaystyle HO-\overset{O}{\overset{\|}{C}}-\overset{O}{\overset{\|}{C}}-OH}{}$$

d. $CH_3\underset{\overset{|}{OH}}{CH}CH_2CH_2CH_2CH_3$

e.
$$CH_3CH_2\overset{O}{\overset{\|}{C}}\!-OH$$

f.
$$CH_3\overset{O}{\overset{\|}{C}}\!-O-CH_2CH_3$$

g.
$$CH_3CH_2\overset{O}{\overset{\|}{C}}CH_3$$

h.

7. a. $CH_3CH_2OH \ + \ H_2SO_4 \ \rightarrow \ CH_2{=}CH_2 \ + \ HOH$
(depending upon conditions, the product may be $CH_3CH_2OCH_2CH_3$)

b. $CH_3CH_2\overset{O}{\overset{\|}{C}}H \ \overset{Na_2Cr_2O_7}{\longrightarrow} \ CH_3CH_2\overset{O}{\overset{\|}{C}}\!-OH$

c. $CH_3CH_2\underset{\overset{|}{OH}}{C}HCH_3 \ \overset{KMnO_4}{\longrightarrow} \ CH_3CH_2\overset{O}{\overset{\|}{C}}CH_3$

d. $CH_3CH{=}CHCH_2CH_3 + H_2(g) \ \overset{\text{metal catalyst}}{\longrightarrow} \ CH_3CH_2CH_2CH_2CH_3$

e. $CH_2{=}CHCH_2CH_3 \ + \ HCl \ \rightarrow \ CH_3\underset{\overset{|}{Cl}}{C}HCH_2CH_3$

f. $CH_3CH_2\overset{O}{\overset{\|}{C}}\!-OH \ \overset{LiAlH_4}{\longrightarrow} \ CH_3CH_2CH_2OH$

PRACTICE TEST (30 Minutes)

1. Write systematic names for each of the following compounds.

a. $CH_3CH_2-O-CH_2CH_3$ _____

b. CH_3CHCH_3 _____
   |
   CH_3

c. $CH_2=CHCH=CH_2$ _____

d. $CH_3C \equiv CCH_3$ _____

e. CH_2Cl_2 _____

f. CH_3CH_2COOH _____

2. Draw the structure of each of the following compounds:

a. 3-hexene   b. heptanal
c. 1,3-dibromobutane   d. 2-butanone
e. 2-pentanol   f. ethyl butyrate
g. ethanoic acid   h. methylamine

3. According to Markovnikov's Rule, which of the following compounds should result when 1-pentene is treated with HCl?

a. 1-chloropentene   b. 1-chloropentane
c. 2-chloropentene   d. 2-chloropentane

4. Which of the following compounds is expected to form a secondary alcohol when treated with a Grignard reagent?

a. formaldehyde   b. butanone
c. formic acid   d. pentanal

5. When a ketone is reacted with a Grignard reagent, the product is usually a(n) _____ ; when a ketone is reacted with a reducing agent, the product is usually a(n) _____ .

6. (**L7**) Based on your knowledge of Valence Bond Theory, answer the questions below regarding the structure of the pesticide Baygon, which is shown on the right.

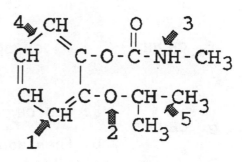

a. Describe the type of hybridization of the carbon atom labeled #1. _____

b. Describe the type of hybridization on the oxygen atom labeled #2. _____

c. Describe the type of hybridization on the nitrogen atom labeled #3. _____

d. Which carbon-carbon bond is shorter, the one labeled #4, the one labeled #5, or are they the same length? _____

7. (**L1 thru L7**) Write the correct product(s) to complete the following equations.

$$
\begin{array}{c}
\quad\quad\quad O \\
\quad\quad\quad \| \\
a. \ CH_3CH_2CCH_3
\end{array}
\xrightarrow{\text{LiAlH}_4}
$$

b. $H_2C=CH_2$ + Br_2 →

c. $CH_3CH_2CH_2-OH$ $\xrightarrow{\text{PCC}}$

d. $CH_3CH_2CH_2-OH$ $\xrightarrow{\text{KMnO}_4}$

e. $CH_3CH-CHCH_2CH_3$ + KOH(ethanol) →
       | |
       H   Cl

$$
\text{f. } CH_3NH_2 \quad + \quad \overset{\overset{\displaystyle O}{\|}}{CH_3CH_2C}-Cl \quad \rightarrow
$$

$$
\text{g. } \overset{\overset{\displaystyle O}{\|}}{CH_3CH} \qquad
\begin{array}{l}
\text{a. add } CH_3CH_2CH_2MgI \\
\quad\quad\quad\quad \rightarrow \\
\text{b. then add water}
\end{array}
$$

8. Which of the following compounds can have cis and trans isomeric forms?

a. $Cl-CH=CH-Cl$ b. $Cl-CH_2-CH_2-Cl$ c. $Cl-C \equiv C-Cl$

9. a. Outline a method for synthesizing 3-pentanol from ethene, propanal, and any necessary inorganic reagents.

b. Outline a method for synthesizing propyl proprionate from 1-propanol and any necessary inorganic reagents.

c. Outline a method for synthesizing butanoic acid from 1-chloro-propane and any necessary inorganic reagents.

CONCEPT TEST ANSWERS

1. D. $CH_2=CH_2$ 2. methane
3. ethyl alcohol or ethanol
4. A. n-Propane   B. n-Butane
   C. n-Hexane   D. Ethane
5. longest continuous   6. A. $LiAlH_4$
7. ethylene glycol
8. steam cracking
9. acetic acid, butyric acid   10. esters
11. saponification   12. amines
13. -al   B. aldehyde
    -ene C. carbon-carbon double bond
    -yne F. carbon-carbon triple bond
    -oic acid E. organic acid
    -ol   D. alcohol
    -one A. ketone
14. ortho, para
15. catalytic cracking, catalytic reforming

1. a. diethyl ether
   b. 2-methylpropane (isobutane)
   c. 1,3-butadiene   d. 2-propyne
   e. dichloromethane   f. propanoic acid

2. a. $CH_3CH_2CH=CHCH_2CH_3$

   b. $CH_3CH_2CH_2CH_2CH_2CH_2\overset{\overset{\displaystyle O}{\|}}{C}H$

   c. $BrCH_2CH_2\underset{\underset{\displaystyle Br}{|}}{C}HCH_3$

   d. $CH_3\overset{\overset{\displaystyle O}{\|}}{C}CH_2CH_3$

   e. $CH_3\underset{\underset{\displaystyle OH}{|}}{C}HCH_2CH_2CH_3$

   f. $CH_3CH_2CH_2\overset{\overset{\displaystyle O}{\|}}{C}OCH_2CH_3$

   g. $CH_3\overset{\overset{\displaystyle O}{\|}}{C}-OH$

   h. CH_3NH_2

3. d. 2-chloropentane

4. d. pentanal

5. tertiary alcohol, secondary alcohol

6. a. sp^2 b. sp^3 c. sp^3
(Note on answer c: The actual hybridization on this type of nitrogen is most likely sp^2, but this wouldn't be predicted based on what you know.)
d. bond #4 is longer; the carbon-carbon bond in the benzene ring is not really a single bond, but has a bond order of about 1.5.

7.
a. $CH_3CH_2\overset{\overset{\displaystyle O}{\|}}{C}CH_3$ $\xrightarrow{\text{LiAlH}_4}$ $CH_3CH_2\underset{\underset{\displaystyle OH}{|}}{C}HCH_3$

b. $H_2C=CH_2$ + Br_2 $\rightarrow$ $BrCH_2CH_2Br$

c. $CH_3CH_2CH_2-OH$ $\xrightarrow{\text{PCC}}$ $CH_3CH_2\overset{\overset{\displaystyle O}{\|}}{C}-H$

417

d. $CH_3CH_2CH_2-OH$ $\xrightarrow{KMnO_4}$ $CH_3CH_2\overset{\overset{O}{\|}}{C}-OH$

e. $CH_3\underset{\underset{H}{|}}{CH}-\underset{\underset{Cl}{|}}{CH}CH_2CH_3$ +KOH(ethanol) $\rightarrow$ $CH_3CH=CHCH_2CH_3$

f. $CH_3NH_2 + CH_3CH_2\overset{\overset{O}{\|}}{C}-Cl \rightarrow CH_3CH_2\overset{\overset{O}{\|}}{C}-NH-CH_3$ + HCl

g. $CH_3\overset{\overset{O}{\|}}{CH}$
  a. $CH_3CH_2CH_2MgI$
  $\xrightarrow{}$
  b. then add water
  $CH_3-\underset{\underset{CH_2CH_2CH_3}{|}}{\overset{\overset{OH}{|}}{C}}-H$

8. A. Cl-CH=CH-Cl, There would be free rotation around the carbon-carbon bond in the second compound. There is no rotation around the triple bond, and the molecule is linear.

9. a. React the ethene with hydrogen chloride to form the ethyl chloride, then combine this with magnesium to produce the Grignard reagent, CH_3CH_2MgCl. React this substance with the propanal and then add water to form the desired alcohol.

b. First oxidize part of the total amount of 1-propanol available to propanoic acid using sodium dichromate or other appropriate oxidizing agent. Then heat the propanoic acid with the remainder of the 1-propanol in acidic solution to form the ester, which is the desired product.

c. First, add magnesium to the 1-chloropropane, then combine the resulting Grignard reagent with carbon dioxide, and finally add water.

CHAPTER 27
POLYMERS: NATURAL AND
SYNTHETIC MACROMOLECULES

LEARNING GOALS:

1. Be familiar with some of the most important natural polymers, including proteins and carbohydrates and know what simple molecular components, such as amino acids, are the building blocks for these natural macromolecules. Nucleic acids and nucleotides play an important role in the chemistry of living systems and so should be given special attention. (Sec. 27.2, 27.3, & 27.4)

2. Understand some of the ways in which synthetic polymers are used in our modern society, how they are classified, and some of the common methods by which they are produced and characterized. (Sec. 27.5)

IMPORTANT NEW TERMS:

acrylics (24.5)
addition polymer (27.0)
adenosine triphosphate (ATP) (27.4)
amino acid (27.1)
carbohydrate (27.3)
cellulose (27.3)
chiral center (27.1)
condensation polymer (27.0)
conjugated protein (27.2)
copolymers (27.5)
deoxyribonucleic acid(DNA) (27.4)
disaccharide (27.3)
elastomer (27.5)
fiber (27.5)
fibrous protein (27.2)
globular protein (27.2)
glucoside linkage (27.3)
glycogen (27.3)
macromolecule (27.0)

monomer (27.0)
monosaccharide (27.3)
nucleic acid (27.4)
nylon (24.5)
peptide bond (27.2)
plastic (27.5)
polyester (24.5)
polyethylene (27.5)
polymer (24.0)
polysaccharide (27.3)
primary protein structure (27.2)
protein (27.2)
ribonucleic acid(RNA) (27.4)
rubber (27.5)
secondary protein structure (27.2)
simple protein (27.2)
starch (27.3)
thermoplastic (27.5)
thermosetting (27.5)
zwitterion (27.1)

CONCEPT TEST

1. If a polymer has the same empirical formula as the monomeric component, it is called a(n)

_____ polymer.   If the polymer is formed by eliminating a small molecule, such as water, from the monomeric combination, it is called a(n) _____ polymer.

2. _____ is the name given to the type of compound formed in plants by the photosynthetic reaction of carbon dioxide, water, and energy from the sun.

3.   When monosaccharides condense, with the elimination of water, the products, such as sucrose, are called _____.

4. _____ is one example of an important polysaccharide.

5. _____, which controls the synthesis of RNA, is the permanent repository of genetic information in the nucleus of the cell.

6. A critical point of the Watson-Crick model of the double helix of DNA is that the base unit of one strand must be paired with a complementary partner on the other strand. For example, adenine is always paired with _____, and guanine is always paired with _____.

7. Polymers are classified on the basis of their response to heating. The type of polymers that soften upon heating and are not altered chemically are called _____. Those that degrade or decompose on heating are called _____.

8. _____ is a polysaccharide found in cotton, the woody part of trees, and the supporting material of plants and leaves.

9. Since humans cannot digest cellulose, _____ is the polysaccharide that serves as our major source of glucose.

10. Synthetic polymers are arranged into three classes depending upon their elasticity; those that can be stretched the most are called _____ and those that are least elastic are called _____.

11. Natural rubber was too brittle, soft, and tacky to be useful until the American inventor, Charles Goodyear, discovered a way to improve the properties by a treatment called _____.

12. The most commonly used polymer is _____.

13. _____, a polymer of 2-methyl-1,3-

butadiene, is often called an elastomer, even though it is obtained from a tree and so is not a synthetic polymer.

14. _____ is the first fiber to result from a deliberate search for a purely synthetic material.

15. Give at least one example of a synthetic fiber that is widely used. _____

STUDY HINTS:

1. The repeating unit in a polymer is not just the simplest unit that is repeated, but rather is determined by the monomer (or monomers) linked to create the polymer structure. For example, the repeating unit in polyethylene, shown below in brackets, is not CH_2 but rather CH_2CH_2, since that is the structure of the monomer, ethylene.

$$- CH_2 - CH_2 [- CH_2 - CH_2 -] CH_2 -$$

2. Sometimes it's difficult to distinguish between addition and condensation polymerization based only on the formula of the repeating unit. Compare the polymerization reactions of ethylene glycol and ethylene oxide shown below.

ethylene glycol

$$HO - CH_2 - CH_2 - OH \rightarrow [- O - CH_2 - CH_2 - O -]$$
$$(-H_2O)$$

ethylene oxide

$$\overset{O}{\underset{CH_2 - CH_2}{\diagup \diagdown}} \rightarrow [- O - CH_2 - CH_2 - O -]$$

The product appears to be the same in both cases, even though the first reaction is clearly a condensation mechanism and the second is an additive mechanism. (Notice, however, that polymers having the same repeating unit will usually have different properties if they have been formed by different reactions.)

3. Some of the methods used in this chapter to

determine the average molecular masses of polymers, such as freezing-point depression and boiling-point elevation, are usually effective only up to molar masses of approximately 40,000 g/mol. Since many polymers have molar masses far in excess of this limit, other methods not discussed in the textbook are used for these determinations.

4. Stereochemistry is another topic which is best studied with the use of molecular models. If models are available, be sure to use them to help clarify this material.

PRACTICE PROBLEMS

1.(**L1 & L2**) Which of the following substances are polymeric?

a. wood   b. grass   c. meat
d. cotton   e. rubber tires   f. starch
f. paint   g. silk   h. proteins
i. permanent press clothing

2. (**L2**) When discussing polymers it is common to talk about the _average_ molar mass. Why do we emphasize that this is an average value when dealing with polymers?

3. (**L2**) Identify each of these reactions as either an addition polymerization or a condensation polymerization reaction.

a. $CF_2 = CF_2 \rightarrow [- CF_2 - CF_2 -]$

b. $CH_2 = CH - CH = CH_2 \rightarrow [- CH_2 - CH = CH - CH_2 -]$

c. $HO-CH_2 - \langle \bigcirc \rangle - CO_2H \rightarrow [- O-CH_2 - \langle \bigcirc \rangle - \overset{O}{\overset{\|}{C}} -]$

4. (L2) Osmotic pressure measurements are a common method for determining the average molar mass of polymers, especially for large molecules. In cases where there is little interaction between the polymer and the solvent, it is possible to use the equation for osmotic pressure discussed in Chapter 14 of the text. At $25^{\circ}C$, an osmotic pressure of 0.0021 atmospheres results when 2.00 grams of polyisobutylene is dissolved in 100.0 mL of benzene. Calculate the average molar mass of polyisobutylene.

5. (L1) A chemist is studying one of the natural macromolecules that is thought to be important in nitrogen fixation and suspects that the sulfur atoms in the molecule play an important part in this process. If the molar mass of the compound is 72,000 and the compound is 0.268% sulfur by weight, how many sulfur atoms are present in each molecule of this macromolecular compound?

6. (L2) Place a bracket around the repeating unit for each of the following simple addition polymers.

a.
$$
\begin{array}{ccc}
CH_3 & CH_3 & CH_3 \\
| & | & | \\
-CH_2-C-CH_2-C-CH_2-C- \\
| & | & | \\
CH_3 & CH_3 & CH_3
\end{array}
$$

b.
$$
\begin{array}{ccc}
CN & CN & CN \\
| & | & | \\
-CH_2-CH-CH_2-CH-CH_2-CH-
\end{array}
$$

c.
$$
\begin{array}{cccc}
Cl & Cl & Cl & Cl \\
| & | & | & | \\
-CH_2-C-CH_2-C-CH_2-C-CH_2-C- \\
| & | & | & | \\
Cl & Cl & Cl & Cl
\end{array}
$$

d.
$$
\begin{array}{cccc}
CH_3 & CH_3 & CH_3 & CH_3 \\
| & | & | & | \\
-CH-O-CH-O-CH-O-CH-O
\end{array}
$$

7. (**L1**) Which is more basic, a nucleoside or a nucleotide?

8. (**L1**) If the sequence on a section of one chain of a double helix is ATGACGTAAT, what is the sequence on the adjacent chain in this section of the double helix?

9. (**L2**) What is vulcanization and why is it useful?

10. (**L1**) For the compounds below, place an asterisk on each carbon atom that is a chiral center.

```
a.     Cl   F              b.            H    H
       |    |                            |    |
CH₃ -  C -  C - Br            CH₃ -  C -  C -COOH
       |    |                            |    |
      Br   Cl                            H   NH₂
```

PRACTICE PROBLEM SOLUTIONS

1. Surprising as it may seem, every one of these substances is a natural or artificial polymer.

2. Polymerization reactions lead to polymer chains with varying numbers of units, and so the molar mass of each individual unit may vary considerably. The average molar mass indicates the average length of the polymer chains, making it a helpful indication of the actual properties of the resulting material.

3. Reactions a and b are simple addition, since there is no removal of a water molecule. In reaction c, however, a water molecule is removed, so this is a condensation reaction.

4.
$$\Pi = \frac{mRT}{V \times \text{ave. molar mass}}$$

Rearranging and inserting known values

$$\text{ave. molar mass} = \frac{2.00\text{g} \times 0.08205 \text{ Latm/mole K} \times 298 \text{ K}}{0.100 \text{ L} \times 0.0021 \text{ atm}}$$

ave. molar mass = 2.3 x 10^5 g/mole

5. First, find out how many grams of sulfur are in one mole of the compound

gram of S/mole of cpd
= 0.00268 x 72,000 g/mole = 192.96 grams

Next, find out how many moles of sulfur are in one mole of compound

$$\text{mole of S/mole of cpd} = \frac{192.96 \text{ g of S/mol of cpd}}{32.1 \text{ g of S/mole of S}}$$

mole of S/mole of cpd = 6 mole of S/mole of cpd

Six moles of sulfur atoms in one mole of compound indicates <u>six sulfur atoms per molecule.</u>

6. a.
$$-CH_2-C\begin{array}{c}CH_3\\|\\\\|\\CH_3\end{array}\left[-CH_2-C\begin{array}{c}CH_3\\|\\\\|\\CH_3\end{array}\right]CH_2-C\begin{array}{c}CH_3\\|\\-\\|\\CH_3\end{array}$$

b.
$$-CH_2\left[-CH\begin{array}{c}CN\\|\\\\\end{array}-CH_2\right]-CH\begin{array}{c}CN\\|\\\\\end{array}-CH_2-CH\begin{array}{c}CN\\|\\\\\end{array}-$$

c.
$$-CH_2-C\begin{array}{c}Cl\\|\\\\|\\Cl\end{array}\left[-CH_2-C\begin{array}{c}Cl\\|\\\\|\\Cl\end{array}-\right]CH_2-C\begin{array}{c}Cl\\|\\\\|\\Cl\end{array}-CH_2-C\begin{array}{c}Cl\\|\\\\|\\Cl\end{array}-$$

d.
$$-CH-O\begin{array}{c}CH_3\\|\\\\\end{array}\left[-CH-O-\begin{array}{c}CH_3\\|\\\\\end{array}\right]CH-O-\begin{array}{c}CH_3\\|\\\\\end{array}$$

7. The phosphate ester in the nucleotide structure will be less basic.

8. Remember that the only possible pairings are adenine with thymine and guanine with cytosine, that is, A with T and C with G. Therefore the structure will be TACTGCATTA.

9. When natural or synthetic rubber is heated with sulfur, the individual polymer chains are linked together with sulfur bonds. This produces a rubber that is more useful commercially because it's harder, stronger, and less thermoplastic.

10. a.

$$CH_3 - \overset{\overset{Cl}{|}}{\underset{\underset{Br}{|}}{C^*}} - \overset{\overset{F}{|}}{\underset{\underset{Cl}{|}}{C^*}} - Br$$

b.

$$CH_3 - \overset{\overset{H}{|}}{\underset{\underset{H}{|}}{C}} - \overset{\overset{H}{|}}{\underset{\underset{NH_2}{|}}{C^*}} - COOH$$

PRACTICE TEST (40 Min.)

1. Name the five organic bases that form nucleic acids and nucleotides.

2. DNA is composed of two polymeric strands linked together by hydrogen bonds to form a coiled structure called a(n) _____.

3. Synthetic polymers that degrade or decompose upon heating are classified as _____ polymers.

4. Name several polymeric substances that are commonly encountered in everyday life.

5. Both natural rubber and gutta-percha are polymeric forms of the same compound, 2-methyl-1,3-butadiene and yet these materials have very different properties. State how these properties differ, and explain these differences in terms of the molecular structure of the polymer.

6. Cellulose is a polysaccharide that is the most abundant organic compound on earth, yet humans cannot use it for food. Explain this observation?

7. If the sequence on a section of one chain of a double helix is TGCATTACGT, what is the sequence on the adjacent chain in this section of the double helix?

8. Benzoyl peroxide can be used as an initiator, that is it will break down to produce species that

427

will start the polymerization process. The half-life of benzoyl peroxide at 70°C in one such experiment is 7.3 hours. Calculate the fraction of the benzoyl peroxide that will remain after this reaction has been allowed to continue for 12 hours.

9. Hemoglobin, a natural macromolecule, is responsible for oxygen transport in human blood. Oxygen gas is bonded to iron atoms in the hemoglobin. If the approximate molecular mass of hemoglobin is 68,000 g/mole, and this substance is 0.33% iron by weight, how many iron atoms are in each molecule of hemoglobin?

10. Place a bracket around the repeating unit for each of the following simple addition polymers.

a.

$-S -$ ⬡ $-S-$ ⬡

b.

$-CH_2-O-CH_2-O-CH_2-O-$

c.

```
    F F F F F F F
    | | | | | | |
   -C-C-C-C-C-C-C-
    | | | | | | |
    F F F F F F F
```

d.

```
        CH₃      CH₃       CH₃
         |        |         |
  -CH₂ -CH-CH₂ -CH-CH₂-CH-
```

11. For the compounds below, place an asterisk on each carbon atom that is a chiral center.

a.
```
       H
       |
       C = O
       |
   H - C - OH
       |
     H₂C-OH
```

b.
```
       OH
       |
       C = O
       |
       C = O
       |
       OH
```

c.
```
        OH
        |
        C = O
        |
  H₂N - C - H
        |
        CH₃
```

12. Analysis of a certain polymer sample has indicated that it consists of components with approximately the following contributions: a mole fraction of 0.3 has a molecular mass of 90,000 g/mole, a mole fraction of 0.6 has a molecular mass of 130,000 g/mole, and a mole fraction of 0.1 has a

molecular mass of 150,000 g/mole.   What is the
average molar mass of this polymer sample?

CONCEPT TEST ANSWERS

1. addition, condensation
2. carbohydrates   3. disaccharides
4. cellulose, starch, or glycogen are acceptable
5. DNA, or Deoxyribonucleic acid
6. thymine, cytosine
7. thermoplastic, thermosetting

8. cellulose   9. starch
10. elastomers, fibers   11. vulcanization
12. polyethylene   13. rubber
14. nylon
15. nylon, dacron, and orlon are all mentioned in
    the text.

PRACTICE TEST ANSWERS

1. (**L1**) adenine, guanine, thymine, cytosine, and
   uracil

2. (**L1**) double helix

3. (**L2**) thermosetting

4. (**L2**) There are many possible answers to this
   question. The textbook and the answer to
   question 1 in the practice problems will provide
   a large number of the chemicals normally
   encountered in everyday life that are polymers.

5. (**L2**) Natural rubber is an elastomer, but gutta-
   percha is brittle, hard, and nonelastic. The
   difference is that the substituent groups lie on
   the same side of the C=C double bond in natural
   rubber, but on opposite sides in gutta-percha.
   The latter structure allows the chains to fit
   together and produces an inflexible structure.

6. (**L1**) Humans cannot digest cellulose because they
   don't have the correct enzyme to catalyze this
   reaction. Cellulose can be used as a food
   source by a number of species, including

termites, cockroaches, cows, sheep, goats, and camels.

7. (**L1**) ACGTAATGCA

8. (**L2**) fraction benzoyl peroxide remaining = 0.32

9. (**L1**) four iron atoms per molecule

10. (**L2**)

a.
$$-\left[S - \bigcirc\!\!\!\!\bigcirc - S - \bigcirc \right]-$$

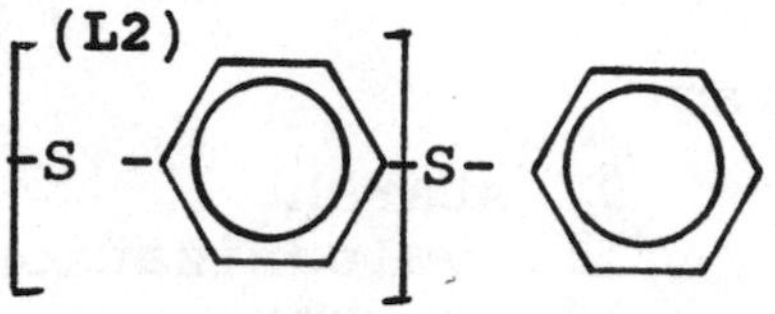

b.
$$-CH_2 \left[O - CH_2 \right] O - CH_2 - O -$$

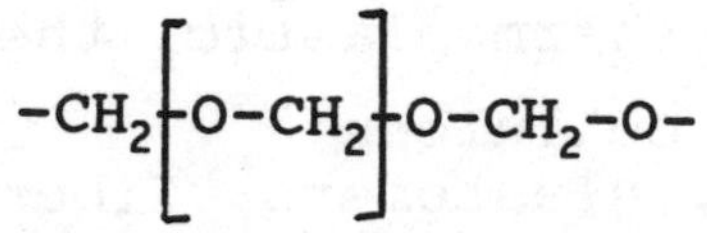

c.
$$\begin{array}{ccccccc} F & F & F & F & F & F & F \\ | & | & | & | & | & | & | \\ -C & -C & -C & -C & -C & -C & -C- \\ | & | & | & | & | & | & | \\ F & F & F & F & F & F & F \end{array}$$

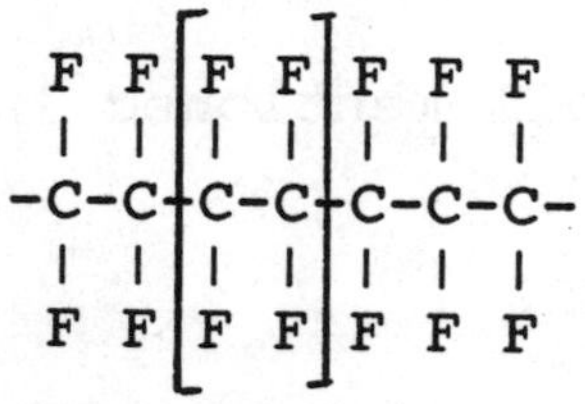

d.
$$CH_2 \left[\begin{array}{c} CH_3 \\ | \\ CH - CH_2 \end{array} \right] \begin{array}{c} CH_3 \\ | \\ CH - CH_2 \end{array} \begin{array}{c} CH_3 \\ | \\ CH \end{array}$$

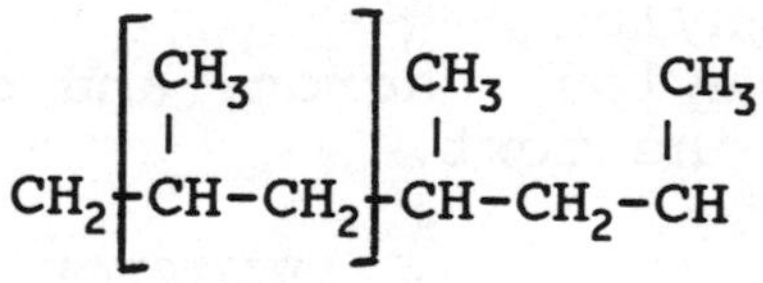

11. (**L1**)

a.
$$\begin{array}{c} H \\ | \\ C = O \\ | \\ H - C^* - OH \\ | \\ H_2C - OH \end{array}$$

b.
$$\begin{array}{c} OH \\ | \\ C = O \\ | \\ C = O \\ | \\ OH \end{array}$$

c.
$$\begin{array}{c} OH \\ | \\ C = O \\ | \\ H_2N - C^* - H \\ | \\ CH_3 \end{array}$$

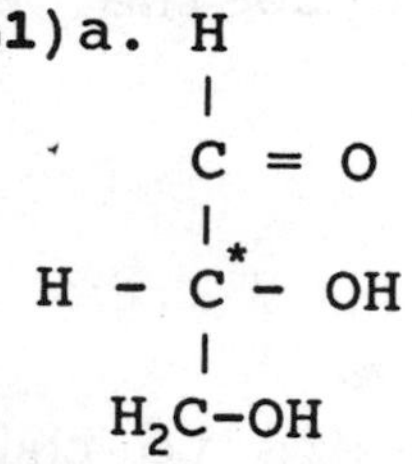

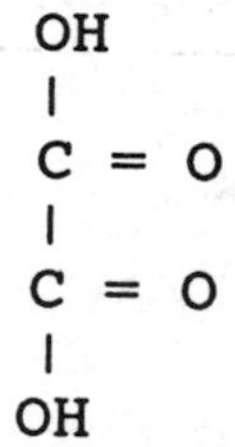

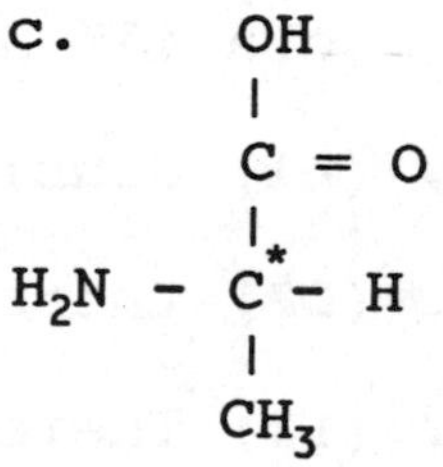

12. (**L2**) 120,000 g/mole